$$\frac{126 - \text{Mono} - 906 \; ①}{320}$$

BRITISH PROSOBRANCH
AND OTHER OPERCULATE
GASTROPOD MOLLUSCS

A NEW SERIES

Synopses of the British Fauna

No. 2

BRITISH PROSOBRANCH AND OTHER OPERCULATE GASTROPOD MOLLUSCS

Keys and Notes for the Identification of the Species

ALASTAIR GRAHAM

Department of Zoology, University of Reading, Reading, England

1971
Published for
THE LINNEAN SOCIETY OF LONDON
by
ACADEMIC PRESS
LONDON AND NEW YORK

ACADEMIC PRESS INC. (LONDON) LTD
24–28 Oval Road
London, NW1 7DX

U.S. Edition published by
ACADEMIC PRESS INC.
111 Fifth Avenue,
New York, New York 10003

Library of Congress Catalog Card Number: 78–141732
ISBN: 0-12-294850-5

Printed in Great Britain by
Butler and Tanner Ltd, Frome and London

Foreword

British Prosobranchs is No. 2 of the New Series of *Synopses of the British Fauna* to be published by the Linnean Society.

The *Synopses* are working field and laboratory books designed to meet the needs of amateur naturalists, sixth-form pupils and undergraduates, though they should also be of value to professional zoologists. The *Synopses* are written by specialists in their respective fields.

The format of the New Series has been changed markedly from that of the original synopses produced by the Society between the Wars and immediately after the Second World War. In the New Series it is hoped that there will be greater uniformity throughout than existed before.

The cover of each *Synopsis* is water-proofed and should withstand quite damp conditions, though total immersion in a rock pool is not advocated. Spaces have been purposely left throughout the text for the owner's field notes and recordings, though unfortunately these spaces have had to be restricted in some places to prevent the *Synopses* exceeding pocket-size dimensions.

The Linnean Society is anxious that the *Synopses* should fill the large gap between the more popular field texts and specialist monographs, and the Society welcomes suggestions and comments so that the series maintains a high standard of content coupled with the maximum degree of usefulness in the field and laboratory.

Already published:
  No. 1. *British Ascidians* by R. H. Millar.
In preparation:
  No. 3. *British Marine Isopods* by E. Naylor.

DORIS M. KERMACK,
Editorial Secretary, Linnean Society

A Synopsis of the British Prosobranch and other Operculate Gastropod Molluscs

ALASTAIR GRAHAM

Department of Zoology, University of Reading, Reading, England

CONTENTS

Introduction

There are three keys on the following pages; one for animals found in marine habitats, the second for those met with in fresh water and a third for the two terrestrial operculate gastropods found in Britain. Estuarine animals may be identified predominantly by way of the marine key, but many can be run down from the freshwater one, so that it is hoped that the fact that there is no hard and fast line separating brackish and truly freshwater habitats may be overcome.

Each key leads to the identification of the family to which the specimen belongs and a reference to the page on which that family is dealt with. Here important familial characters are summarized, and further keys, if required, lead to the identification of, first, the genus and then the species to which the specimen belongs. Lastly, a few notes on each species are given. These include its broad distribution in this country and the height of the shell in millimetres, except in the case of limpets, where the measurement refers to the anteroposterior length. In most cases I have tried to use the true generic and specific characters in the keys except where these are based on internal features, since my aim has been to allow identification on external features and leave the animal alive and intact for further work. The use of critical specific features has not always been possible, and in some groups (e.g. pyramidellids, which are notoriously difficult to identify; and rissoids) recourse has to be had to probably unimportant details. For this reason drawings are given of most shells, since it is too difficult a task to try to describe in words the elaborate three-dimensional shape of the gastropod shell. In most cases shells are drawn with the apex upwards and the mouth towards the observer. The scale by each figure gives the height in millimetres. In view of the small number of terrestrial species it was not thought necessary to figure them, nor to give figures of most of the freshwater species in view of the excellent drawings in the key to freshwater gastropods in the series published by the Freshwater Biological Association. Some of the drawings illustrating this *Synopsis* are taken by kind permission of the Council of the Ray Society, from *British Prosobranch Molluscs*, by V. Fretter and A. Graham.

General Organization

The animals which are dealt with in this *Synopsis* do not constitute a taxonomically unified group of molluscs. One of the features which characterizes most prosobranch gastropods, the operculum, is also found in some species of opisthobranchs. The ordinary collector meeting an unfamiliar gastropod with an operculum would inevitably, and, in most cases, quite properly believe that he had found a prosobranch. It seemed more helpful, therefore, to extend the *Synopsis* a little so as to cover not only the prosobranch gastropods, but also that small number of others which are both operculate and opisthobranch.

Gastropods are the animals commonly called snails and slugs. Most are marine, the habitat in which the group evolved, but gastropods have also invaded fresh water and some members of a very successful group, the pulmonates, have become terrestrial and breathe air. Terrestrial gastropods have never acquired the ability to control water loss through their skin, however, and are therefore forced into inactivity unless the environmental humidity is high.

The gastropod body consists of a ventral sensory and locomotor section, the *head-foot*, and a dorsal part containing the viscera, the *visceral mass* or *visceral hump*. On the head are placed the *mouth*, the *eyes* and a pair of sensory *tentacles*; the foot is a broad muscular organ with a flattened sole lubricated by mucous secretions on which the animal crawls, and the visceral hump is characteristically wound in a right-handed spiral, lodges the viscera and bears an enveloping calcareous *shell* of the same shape, secreted by the skin on its surface which is known as the *mantle*. In the ancestors of gastropods the posterior surface of the visceral mass had a pit-like depression under the shell, the *mantle cavity*, roofed by an extension of the mantle known as the *mantle skirt*, in which lay the animal's *gills* and into which there discharged the *anus* and the united *excretory* and, *genital ducts*. The head-foot could be withdrawn into this—not necessarily wholly—and so into the shelter of the shell, if danger threatened, and extended later: during this activity the mantle cavity acted as a compensation sac, filling with water on protrusion of the head-foot and emptying when it was withdrawn. The retraction was accomplished by shell muscles, of which there were two, running from the shell to the right and left sides of head and foot.

All living gastropods have been affected by torsion, a process in which the visceral hump is rotated anticlockwise on the head-foot. It has the effect of bringing the originally backward-facing mantle cavity forwards over the head, and makes its original right side lie on the left and the original left face right. All structures which link any part of the body in the head to any part in the visceral hump now cross the mid-line of the body where head-foot and visceral hump join. The effect of this is to twist the oesophageal region of the gut so that its right side faces left and vice versa, and to introduce a figure-of-eight shape into the nervous system, a condition known as streptoneury. The animals must obviously have gained some considerable advantage to make such a radical transformation of the body worthwhile; two advantages are supposed to derive from torsion: (1) the withdrawal of the more important head before the less important foot into the mantle cavity on disturbance, and, with the developmen of an operculum on the foot, the possibility of total protection against predato or adversity; and (2) the better ventilation of the cavity when its mouth face the direction in which the animal is moving.

At this point in gastropod evolution the animals exhibit right and left sets o paired organs, an anterior mantle cavity and, commonly, a spirally wound shell This stage, the lowest in the evolution of the class, is represented today by th more primitive members of a grade known as the diotocardian or archaeo gastropod: they fall into a group known as the Pleurotomariacea (sometimes a Zeugobranchia). The bodily organization and general metabolic efficiency o these animals is poor, partly because of inadequate ventilation of the mantle cavity, which involves an inhalant current entering from both right and left an escaping in the mid-line by way of a notch or hole in the mantle skirt and shell Because of the spiral coiling of the shell entry of water from the right is restricte and the median outlet is not always good. Two successful evolutionary escape from this situation have been made. In the first the mantle cavity has become disused as a gill chamber and respiratory exchange becomes centred on secondary gills developed on the mantle edge round the mouth of the shell. At the same time the spiral part of the shell is lost and the body of the animal lies only in the expanded part near the mouth. The shell has also re-shaped itself to form a stream lined cone, and the shell muscles unite to a single one with a horseshoe-shaped attachment to the shell. These characters mark a higher level of the diotocardian grade, the extremely successful limpets, placed in the group Patellacea.

The second solution to the problem of adequate ventilation of the mantle cavity has led to a much more plastic stock, members of which have adaptively radiated into an extraordinary variety of ecological niches. Because of the fact that the spiral in which the body and shell lie is dextrally coiled its right side is shorter than its left; there is therefore much less space on that side for the growth of the right members of paired structures. In the course of evolution they have disappeared and only the left member of the pair persists, with greater space in which to grow. In the case of the organs in the mantle cavity the right gill vanishes, its associated sense organ (the *osphradium*, testing the water flowing over the gill) also goes, and the right auricle of the heart, to which drained the blood oxygenated in the right gill, is lost. The right kidney ceases to function as an excretory organ, but part of it is retained because it had provided a way out for the gametes. The primary effect of these changes is that water enters the mantle cavity on the left, sweeps over an expanded left gill and escapes on the right at a point close to where the anus and reproductive opening lie, so that faeces and gametes are swept out in the exhalant stream.

The higher grade of organization and activity which is permitted by these changes is known as the monotocardian or mesogastropod. It has been achieved more than once, so that the grade is polyphyletic. It has also nearly been achieved by a group of diotocardians, the Trochacea or top shells, which have lost the right gill and osphradium, though they retain the right kidney as an excretory organ and the right auricle. It has also been attained (in a very distinct way) by the Neritacea, whose anatomy differs from that of other monotocardians in a

ariety of important details. This group is not well known in this country where
nly one species occurs, the freshwater *Theodoxus fluviatilis*.

Mesogastropods have radiated into many ways of life. Their original habitat
, the marine littoral, and this is where most prosobranchs are still to be found.
'hey may be browsers on weeds or on the film of algae and vegetable detritus
/hich covers the surface of the shore—this is where lacunids, littorinids and
issoids abound; they may specialize in feeding on the fine weeds of rock pools,
ike the minute summer-time rissoellids, omalogyrids and skeneopsids; some
ave become carnivores, browsing on the flesh of sessile animals like sponges
nd tunicates (eratoids, cerithiids, cerithiopsids, triphorids, lamellariids); other
arnivores (naticids) bore holes in the shell of their bivalve prey to get at the
lesh within; some have become dependent on collection of food by ciliary
neans and have enlarged mantle cavity and gill to produce a more powerful
urrent and extensive sieve (calyptraeids), and still others seem to have become
emiparasitic (capulids).

In addition to those mesogastropods which live between tide marks some are
vholly sublittoral, where they occupy corresponding niches in the sublittoral
:ommunity: turritellids are ciliary feeders; scalids and stiliferids are semi-
)arasitic; cypraeids browse on sessile coelenterates. A small number of mesogastropods have evolved the ability to survive in brackish and even fresh water.
In the British fauna this group is restricted to a few rather primitive types,
valvatids, neritids, hydrobiids, bithyniids and viviparids, all vegetarian, and, as
might be expected of animals enclosed in a calcareous shell, rather more at
home in hard than in soft waters. Two species are terrestrial.

The prosobranch gastropods are normally classified so as to include a third
grade known as the stenoglossans or neogastropods. The differences between
these and the mesogastropods are not great and the animals do not exhibit a gain
in organizational efficiency comparable with the step from diotocardian to
monotocardian. The neogastropods are a uniform group showing much mosaic
evolution, so that their proper classification is still doubtful. They are all carnivores, though some eat carrion rather than attack live prey: all have a proboscis
and lay their eggs in capsules fastened to the substratum on which they live. All
are marine.

From some (probably rather lowly) monotocardian stock there sprang the
animals classified in a second major division of the gastropods, the opisthobranchs. The changes involved in this are, in total, vast, so that there is a great
gulf between the more primitive prosobranch and such an advanced opisthobranch as an eolid, devoid of shell, agile in locomotion and with its oddly
arranged alimentary tract. The more primitive opisthobranchs, however, retain
most of the bodily characteristics of their prosobranch ancestors; they have
a spirally coiled shell into which they can retreat, an operculum on their foot
and, internally, a streptoneurous nervous system. Superficially, therefore, they
resemble prosobranchs, and for that reason some of them are dealt with in
the following pages. These include the acteonids and the pyramidellids, the
former carnivores, the latter ectoparasitic.

In order to help in the use of the key and in the understanding of the diagnoses it is necessary to explain a certain number of technical terms. These are incorporated into the following description, where all those that are employed in the keys appear in **bold-faced** type.

The prosobranch body is normally enclosed in a spirally wound shell (Fig. 1) coiling **dextrally,** that is, a clockwise-going spiral when the shell is viewed from above the origin of the spiral or **apex.** The shell opens at the **mouth** or **aperture,** and when the shell is held with the apex uppermost and the mouth facing the observer, the mouth lies right of the axis round which the shell spirals, or **columella.** If the shell is **sinistral,** the mouth lies left of the columella. The columella may be hollow, and the opening at the base of the shell leading into this space is the **umbilicus.** The right boundary of the mouth (shell held as above) is the **outer lip,** its left the **inner lip;** if these meet to form a continuous boundary to the mouth the resulting structure is termed a **peristome.** The basal turn of the spiral, inwards from the mouth, accommodates the body of the snail, and is for that reason known as the **body whorl** and its broadest part is the **periphery;** the other turns of the spiral (**whorls**), which are nearer the apex, constitute the **spire** according to the spatial relationship of successive whorls to one another the spire may be a flat planispiral like the coil of a catherine wheel (Fig. 59), or a conical helicospiral like the familiar snail shell. Where one whorl meets the next the line of contact is known as a **suture.** Whorls may be flat-sided, in which case their profile and the sutures are in the same plane, giving a strictly conical shell whorls, however, have more frequently a convex profile, in which case they curve inwards, or dip, to the sutures. The profile of the spire of a conical shell may be accurately rectilinear (Fig. 20), but it is sometimes slightly concave, or **coelo conoid** (Fig. 23), sometimes slightly convex, **cyrtoconoid** (Fig. 16). In a few families the body whorl of the shell expands abruptly so as to wrap round and conceal all the older whorls; this is known as a **convolute** shell (Figs 77, 78).

The spiral of the typical gastropod shell is extended by secretion from the edge of the mantle underlying its mouth, and during life calcareous material is added to the mouth at more or less regular intervals, long or short, separated by pauses, until maturity is reached and further growth stops. Each of the pauses usually leaves a slightly irregular mark across the whorl and is known as a **growth line.** Growth lines, indeed, mark the position of the mouth of the shell at earlier times in the life history of each individual. As the shell grows various elements of sculpturing frequently appear on its surface: these invariably indicate some departure from regularity of secretion either in space or time. If the amount of calcareous matter secreted along the lips of the mouth is greater at one place and less at another, ridges, separated by grooves, appear on the shell surface. These follow the spiral curve of the whorls and are known as **spiral lines, striae** or **ridges** according to their size. The amount of calcareous matter secreted at the mouth of the shell may show a temporal fluctuation, being greater at one time than another: this gives rise to thicker shell at one moment and thinner at another, the thicker part running across the whorl from the upper (adapical) side to the lower (abapical) side. This kind of thickening is a **rib,** or, if out-turned, a **varix;** often a rib of this sort lies along and strengthens the outer lip, forming a **labial rib** or **varix.** If ribs and spiral ridges both occur a reticulate pattern or

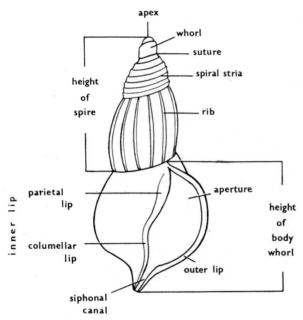

Fig. 1. Diagram of gastropod shell to show nomenclature of parts. Reproduced from *British Prosobranch Molluscs* by Fretter and Graham.

decussation may be formed as the one type of sculpture crosses the other. A ridge or ridges may appear on the columella for the same reason, local increase in rate of secretion. The end of the ridge is seen on looking in at the mouth and resembles a tooth; it is described as a tooth on the columella. The calcareous material out of which the shell is made is embedded in a continuous organic matrix of conchiolin, and the outer layer of calcareous substance lies under a film of the material known as the **periostracum**. It is rarely visible in prosobranch shells, however, though in a few genera, such as *Trichotropis* (Fig. 2), *Capulus* (Fig. 7C) and *Colus* it forms a distinct horny layer over the surface.

When the snail is withdrawn into the shell it is orientated so that the dorsal surface of the head and foot lies under the outer lip, the ventral side faces the columella and inner lip, the left side is basal and the right side lies under the point where the outer lip runs into the body whorl. In a few advanced meso gastropods and in all neogastropods the mantle develops a tubular outgrowth on the left side which forms an inhalant **siphon** (Fig. 3), allowing the animal to draw its respiratory water from some distance from its shell. When this develops, corresponding outgrowth of the shell mouth appears to accommodate and support the siphon; this is the **siphonal canal**, placed where outer lip meets inner. A few neogastropods have a similar break in the circumference of the mouth over the spot where the outgoing current leaves the mantle cavity on the right: this produces an **anal notch** in the outer lip where it joins the body whorl.

The head (Fig. 3) bears the mouth and a pair of **cephalic tentacles**. On the lateral side of each of these is an eye on an **eyestalk.** In some opisthobranchs the eyes retreat into the body and then appear to lie between the tentacles. In many diotocardians the foot also bears a number of tentacles and eye-like structures (Figs 4, 5). The tentacles on the foot are known as **epipodial tentacles** and, in some species, they spring from a continuous fold, set along the side of the foot known as the epipodium. The epipodium may extend on to the head and give rise to lamellar lobes on the neck region and crest-like outgrowths between the cephalic tentacles. The former structures are known as **neck lobes.** Other tentacles may occur on the prosobranch body. Thus many rissoids, lacunids and nassids have one or two tentacular outgrowths at the hind end of the foot or at the point where the operculum is attached: these are **metapodial tentacles** (Figs 3, 6, 7). Rissoids and some other kinds of mesogastropod may also have a tentacle growing from the edge of the mantle skirt on the right side and occasionally also on the left; these are called **pallial tentacles** (Fig. 7). In turritellids a whole series of pallial tentacles lies round the entrance to the mantle cavity.

The mouth lies on the ventral side of the head at the end of a down-curved snout. In some of the higher mesogastropods and in all neogastropods the opening which lies in this position is not, however, the true mouth, but opens into a proboscis sac into which a proboscis is withdrawn when the animal is not feeding. The true mouth lies at the tip of the proboscis. In pyramidellids

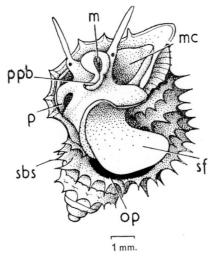

1 mm.

Fig. 2. *Trichotropis borealis*, to show periostracum on shell (*sbs*); *m*, mouth; *mc*, mantle cavity; *op*, operculum; *p*, penis; *ppb*, proboscis; *sf*, sole of foot. From *British Prosobranch Molluscs* by Fretter and Graham.

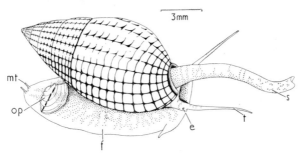

Fig. 3. *Nassarius reticulatus*, to show metapodial tentacles (*mt*). *e*, eye; *f*, foot; *op*, operculum; *s*, siphon; *t*, tentacle. From *British Prosobranch Molluscs* by Fretter and Graham.

a shelf-like projection known as a **mentum** lies ventral to the opening of the proboscis sac, between it and the front edge of the foot. Its importance is unknown.

Within the buccal cavity lies the feeding apparatus of the gastropod, the **odontophore,** carrying the radula along the middle of its dorsal surface. Odontophore and **radula** are partially protruded through the mouth for feeding. The radula consists of a long, cuticular ribbon carrying transverse rows of teeth. The most anterior of these are used in feeding and ultimately get broken off; they are replaced at the same rate by new rows of teeth secreted at the inner end of a blind sac, the radular sac, which opens from the posterior wall of the buccal cavity. The radula is therefore continually moving slowly forwards over the odontophore like a nail over the finger-tip: how this happens is not known.

The radula has some classificatory importance. There are several types, all with the same basic arrangement, which can be correlated with differences in diet. Each transverse row of teeth contains one central tooth, the **rachidian;** this is flanked on each side by a number of **lateral** teeth and these may, in turn, have a number of **marginal** teeth on their outer side. The most primitive radular type (found in Haliotidae, Scissurellidae, Fissurellidae, Trochidae, Turbinidae and Neritidae in the British fauna) is the **rhipidoglossan,** characterized by a very large number of marginals and by five laterals on each side of the rachidian tooth: it is used to brush loose particles off the substratum. The second type, the **taenioglossan** (the common prosobranch type), has lost the large number of marginal teeth and there are seven teeth in each row, three on each side of the rachidian. This radula rasps the substratum as well as brushing it and so can gather particles torn off it as well as those lying upon it. Reduction in the number of teeth per row is carried further in the type of radula known as **rachiglossan** (found in Muricidae, Buccinidae, Nassidae) in which the central tooth is flanked by only one lateral tooth on each side; and in the **toxoglossan** type (Conidae) in which only one tooth per row persists. These two types are both concerned with feeding on animal flesh; the rachiglossan radula allows its owner, at least in certain species, to bore molluscan shells to reach the flesh of the prey, and the toxoglossan teeth, by association with the secretion of a poison gland, form poisoned darts for killing prey.

There are two other radular types which cannot be placed in the same evolutionary sequence as the others: these are the docoglossan and the ptenoglossan. The **docoglossan** occurs in the families Acmaeidae, Patellidae and Lepetidae. The rachidian tooth is small or absent and there are three lateral teeth with hard, black tips, flanked by three smaller, unpigmented marginals on each side. This radula is a powerful rasping tool and the marks which it makes may often be seen on the rocks on which limpets have been browsing. A **ptenoglossan** radula is confined to the families Scalidae and Ianthinidae in the British fauna. Each row in this type contains a large number of similar teeth which have the form of simple, recurved hooks: it forms not so much a rasping as a grasping apparatus and is used by animals which ingest their prey whole, or which hold larger prey whilst parts are bitten off by jaws.

Other openings occur on the foot. A slit across the anterior margin of the foot

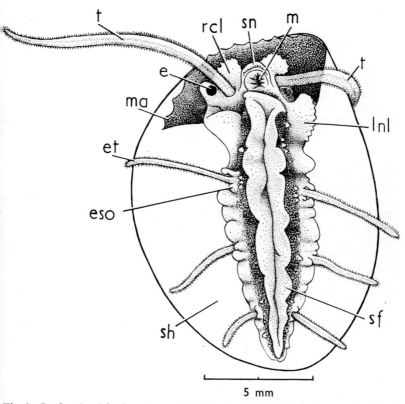

Fig. 4. *Cantharidus clelandi*, to show epipodial tentacles (*et*) and sense organs (*eso*) and neck lobes (*lnl*). *e*, eye; *m*, mouth; *ma*, mantle; *rcl*, right cephalic lappet; *sf*, sole of foot; *sh*, shell; *sn*, snout; *t*, cephalic tentacle. From *British Prosobranch Molluscs* by Fretter and Graham.

B

coincides with the boundary between the anterior part of the foot, the **propodium** (Fig. 72), and the **mesopodium,** which forms the sole on which the mollusc creeps. The propodium attains a large size only in naticids among British molluscs. The slit also acts as the mouth of an **anterior pedal gland** which secretes much of the mucus on which the animal moves. Another source of mucus, the **posterior pedal gland,** forms a prominent median pore in the sole of the foot of many of the small littoral prosobranchs such as rissoids (Fig. 53). It is usually connected to the posterior tip of the foot by a groove, and the mucus passed backwards along this tends to produce a rope on which the animal can clamber from weed to weed or downwards from the surface film of the rock pool in which it lives. Female neogastropods and the females of lamellariids, cypraeids and eratoids have a further aperture in the mid-line of the pedal sole; this is the opening of the **ventral pedal gland,** which is concerned with the laying of egg capsules (Fig. 8). In muricids still another opening in the sole of the foot—in both sexes this time—lies close behind the anterior margin; this houses the **accessory boring organ,** which can be everted and applied to the surface of the shell of the prey as the muricid bores. It secretes a chemical which facilitates and accelerates the boring process.

In diotocardian prosobranchs fertilization is external and the gametes are broadcast, escaping from the body through the opening of the right kidney. In meso- and neo-gastropods and in opisthobranchs a copulatory process is almost universal and the males possess a penis. This (Fig. 26A) is usually formed from part of the foot, arises behind and either above or below the right cephalic tentacle and is kept within the mantle cavity. In viviparids, however, it is the right cephalic tentacle itself which is modified. A few prosobranch males are aphallic: these commonly belong to species with tall, narrow spires to the shells, implying a very narrow mantle cavity, where the presence of a penis might interfere with ventilation, or, as in the case of *Turritella*, with a ciliary food collecting device. In these species the animals either revert to external fertilization, or the females use the ciliary feeding current to draw sperm into the mantle cavity, or the males produce **spermatozeugmata.** These are large cells derived from the same source as spermatozoa but infertile, in which the flagellar apparatus has become hypertrophied to make them capable of ferrying a load of functional sperm from male to female, swimming into the female ducts to release their load.

On the dorsal surface of the foot, near the posterior end, lies the operculum. It is borne on an opercular lobe and is secreted in a relative short transverse groove (Fig. 9). During its growth it undergoes a rotation in a clockwise direction on the opercular lobe, though it is not known how this is achieved, and its size is such that it blocks the mouth of the shell when the animal has withdrawn. In all prosobranchs it is made of conchiolin, but in a few there may be some calcareous material added to the conchiolin, a process which reaches its maximum in the family Turbinidae, the operculum of which is a massive calcareous knob. In the diotocardians (Fig. 10) the operculum (if present) is **polygyrous,** that is composed of many turns; in most monotocardians the number of turns is much reduced and the operculum is then described as **oligogyrous.** In this type the additions during growth may be such as to give it a **spiral** shape or they may take

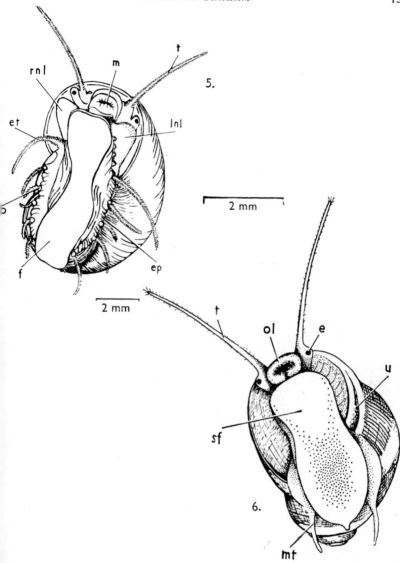

Fig. 5. *Margarites helicinus*, to show epipodium (*ep*), epipodial tentacles (*et*), associated sense organs (*eso*) and neck lobes (*lnl, rnl*). *f*, foot; *m*, mouth. From *British Prosobranch Molluscs* by Fretter and Graham.

Fig. 6. *Lacuna vincta*, to show metapodial tentacles (*mt*). *e*, eye; *ol*, outer lip; *sf*, sole of foot; *t*, cephalic tentacle; *u*, umbilicus of shell. From *British Prosobranch Molluscs* by Fretter and Graham.

the form of **concentric** rings added to the periphery of the older part of the operculum. The central part in each is known as the **nucleus.**

In drawing up the following keys it has been assumed that the reader will have living specimens available for study and the use of a good lens or low power stereo-microscope. A few prosobranchs are shy to emerge from their shells and move around, but the great majority will do this and, unless subjected to sudden jars or vibration will continue to creep about in a lively way. With the help of a mounted needle or similar instrument the shell of the creeping animal may usually be tilted quite readily in one direction or another without causing the retraction of the snail, so that various views of the head, foot, mantle edge, etc may be reasonably easily obtained. These are the only parts of the body to which reference is made in the key and it should therefore be not too difficult to check the necessary point.

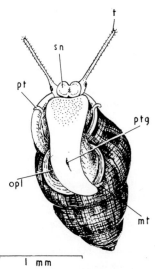

Fig. 7. *Alvania punctura*, to show pallial tentacles (*pt*). *mt*, metapodial tentacle; *opl*, operculigerous lobe; *ptg*, opening of posterior pedal gland; *sn*, snout; *t*, cephalic tentacle. From *British Prosobranch Molluscs* by Fretter and Graham.

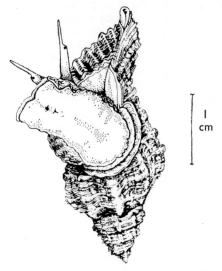

Fig. 8. *Ocenebra erinacea*, to show openings on sole of foot. The animal is female and shows two openings: the anterior is that of the pit in which the accessory boring organ lies, the posterior is that of the ventral pedal gland involved in attaching egg capsules to the substratum and absent in males. From *British Prosobranch Molluscs* by Fretter and Graham.

Collection and Preservation

Marine prosobranchs are most commonly found on boulder beaches and rocky shores, only a few tolerating sand or mud. On rocky shores some, like limpets, are obvious on exposed surfaces, but most tend to shelter against intertidal desiccation by retreating into rock crevices, under stones or into pools.

At the landward edge of a rocky beach *Littorina neritoides* and *L. saxatilis* hide in narrow cracks in the rock. Lower, *L. littorea* occurs, sometimes clinging to rocks but, commonly, in gregarious masses in sheltered recesses of the rock or in rock pools. *L. littoralis* is associated with bladder wracks, especially near the margin of pools. Where barnacles occur small littorinids often hide in empty shells, and dog whelks, *Nucella lapillus*, shelter in groups in crannies or under ledges. Top shells are also to be found on the underside of projecting rocks or in pools. *Tricolia pullus*, many rissoid species and related small prosobranchs occur, particularly in summer, in coralline pools, whilst other species of rissoid prefer slightly silty situations under boulders or in crevices.

Under stones near low water is the place in which are commonly found many species like *Diodora apertura* and *Emarginula reticulata*, *Acmaea virginea* and *A. tessulata*, *Calliostoma zizyphinum*, *Cerithiopsis tubercularis*, *Trivia monacha*, *Ocenebra erinacea* and *Nassarius incrassatus*. *Patina pellucida* is common on *Laminaria* just exposed at low water of spring tides.

Pyramidellids are best found by looking at places where their hosts are occurring in large numbers—large stones on which there is a rich growth of *Pomatoceros*, colonies of *Sabellaria* and mussel beds.

All these animals may be easily picked up by hand or with forceps. Limpets of the genus *Patella* are most successfully dislodged from rocks by a surprise blow with the heel of a Wellington boot; if unsuccessful at the first attempt go on to another animal—a warned limpet increases its grip and can never be dislodged without damage.

On sandy shores *Acteon tornatilis* and *Natica* spp. may be found, both creeping on the surface or slightly under, though both can burrow. On muddy shores, especially in estuarine conditions, *Hydrobia ulvae* abounds: they are most easily collected by skimming the surface of wet places with a plastic tea strainer. Even in this unpromising situation, specimens of edible winkle may be captured.

Take animals home in jars with some weed over them to prevent drying out. Do not cover with water: all intertidal creatures successfully survive periods in air provided that they are kept damp. It is also rewarding to take home handfuls of weed, especially the finer sorts from pools, *Laminaria* holdfasts and the like, and to set them in dishes of sea water: as the water becomes slightly foul small animals—many not otherwise easily seen—creep out and climb to the surface film where they may be collected for examination.

Freshwater animals are most commonly found on weed and are easily collected by gathering that and washing the animals from it. *Viviparus* spp.,

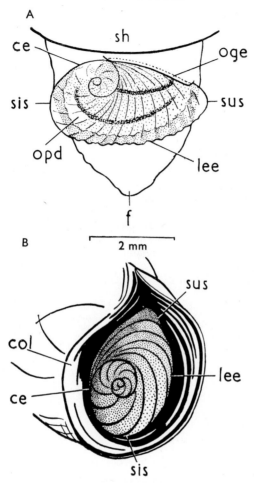

Fig. 9. *Littorina littorea*, to show the relationship of operculum to shell **A,** in a creeping winkle; **B,** in a retracted winkle. *ce,* columellar edge of operculum; *col,* columella of shell; *f,* posterior tip of foot; *lee,* labial edge of operculum; *oge,* opercular groove secreting new opercular material; *opd,* operculigerous disc; *sh,* shell; *sis,* siphonal side of operculum; *sus,* its sutural side. From *British Prosobranch Molluscs* by Fretter and Graham.

however, often burrow into gravel beds, whilst *Potamopyrgus* and *Theodoxus* adhere to stones.

Prosobranchs which remain persistently in their shells may be relaxed. Marine kinds relax best in 7% (= 0·36M) magnesium chloride solution in fresh water or 0·08% nembutal in sea water; freshwater and terrestrial species in 0·08% nembutal in fresh water; soda water is also valuable. It may help if the shell is cracked; this is most effectively and easily done by means of a vice, and the animal can be completely extracted from the remains by scraping the attachment of the shell muscle off the columella. Obviously this method of extracting the animal is nonsensical if one wishes to preserve the shell; if the soft parts are not regarded as important it then suffices to immerse the animal briefly in boiling water when the body can be easily removed with the traditional pin. There is, unfortunately no certain way of extracting the animal from the shell so as to get both parts in perfect condition, though rapid freezing for 6–12h sometimes detaches the columellar muscle and allows extraction of the animal from an intact shell.

If it is desired to preserve animals after relaxing and anaesthetization either formalin (about 10%) or alcohol (70%–90%) may be used. Always remember to use plenty of preserving fluid in relation to the volume of the animal. Shells are best kept dry.

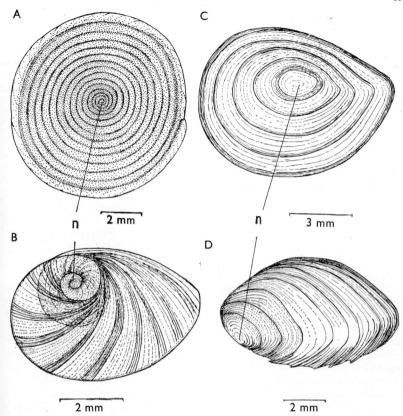

Fig. 10. Types of operculum. **A,** polygyrous spiral, *Gibbula cineraria*; **B,** oligogyrous spiral, *Littorina littorea*; **C,** concentric, *Viviparus viviparus*; **D,** with marginal nucleus, *Nassarius reticulatus*. From *British Prosobranch Molluscs* by Fretter and Graham.

Classification

Any kind of animal which forms a naturally (or potentially) interbreeding population is known to zoologists as a *species*. When different species of animals are examined and their anatomy investigated some are found to be very similar, others less so. Those that are most alike are thought to be so because they are closely related and share a common ancestor, just as brothers and sisters are more like one another than they are like members of the population at large. Similar species can be grouped into units each of which is known as a *genus*, and groups of genera joined into larger assemblages each known as a *family*. In this way a hierarchy of categories is established which can be used to give a classification of any one species. It will be noticed that whereas the idea of a species corresponds to something in nature the other categories depend upon a particular zoologist's impression of what constitutes closeness of relationship. There is therefore no guarantee that any two students of a group of animals will agree in their ideas as to how species should be grouped to form genera, or genera to form families; there is no inevitably *correct* classification, only a generally accepted one. That for the animals included in this *Synopsis* is given below and is based on one drawn up by the German malacologist Johannes Thiele.

In modern classifications the name of a given kind of animal consists of the name of its genus and of the species of that genus to which the animal belongs. The two words are written one after the other like the forename and surname of a man, the genus first, the species second. They are in Latin, and the specific name, if an adjective (as it often is), must agree in number and gender with the generic name. Thus the name of the common limpet is *Patella vulgata*, telling the reader that it is classified in the species *vulgata* of the genus *Patella*; the china limpet is *Patella aspera* and the black-footed *Patella intermedia*. They comprise three different kinds of animals normally breeding only with others of the same species but sufficiently similar in organization to be put into the one genus *Patella* and, in the opinion of most malacologists, sufficiently unlike other limpets (such as those found on weed, which have no gills on the mantle skirt over the head) to require that weed limpets be placed in a different genus, *Patina*.

It is common usage to add to the name of the animal the name of the investigator who first applied that specific name to that particular kind of creature. Thus one often writes *Patella vulgata* L., which tells one that Linnaeus first applied the name *vulgata* to the common limpet of north-west European coasts. Since there have often been instances of the same name being applied to different animals, even of the same genus, it is important to avoid uncertainty as to which animal is being talked of: thus Jeffreys described the china limpet as *Patella depressa* whilst Pennant used the same name for the black-footed limpet. Obviously it would be important to write *Patella depressa* Pennant if one were talking about the latter animal, though both Pennant's and Jeffreys' names for these limpets have been replaced by others.

If a species today falls into the same genus as it was in when first named the name of the author of the specific name is added without parentheses, thus:

Patella aspera Lamarck. In many cases, however, more recent work has shown that it is no longer possible to retain the species within the original genus and a new genus has had to be created, to which the species is now referred. When this has happened it is necessary to place the author's name within parentheses. For example, Linnaeus originally placed the common limpet, which he called *vulgata*, and the blue-rayed limpet, which he called *pellucida*, in the same genus *Patella*. But enough is now known of their anatomy to make most people feel that the inclusion of such unlike animals in one genus makes the concept of the genus too loose. The species *pellucida* is now, therefore, separated in the genus *Patina*. The names are correctly written *Patella vulgata* L. and *Patina pellucida* (L.).

CLASS GASTROPODA
Subclass PROSOBRANCHIA
Order DIOTOCARDIA

Superfamily Zeugobranchia
   Family Haliotidae
      Haliotis tuberculata L.
   Family Scissurellidae
      Scissurella crispata Fleming
   Family Fissurellidae
      Emarginula conica Lamarck
      Emarginula crassa Sowerby
      Emarginula reticulata Sowerby
      Puncturella noachina (L.)
      Diodora apertura (Montagu)

Superfamily Patellacea
   Family Acmaeidae
      Acmaea virginea (Müller)
      Acmaea tessulata (Müller)
   Family Patellidae
      Patella vulgata L.
      Patella aspera Lamarck
      Patella intermedia Jeffreys
      Patina pellucida (L.)
   Family Lepetidae
      Propilidium exiguum (Thompson)
      Lepeta caeca (Müller)
      Lepeta fulva (Müller)

Superfamily Trochacea
   Family Trochidae
      Margarites helicinus (Fabricius)
      Margarites groenlandicus (Gmelin)
      Gibbula magus (L.)

Family Trochidae *cont.*
   Gibbula tumida (Montagu)
   Gibbula umbilicalis (da Costa)
   Gibbula cineraria (L.)
   Gibbula pennanti (Philippi)
   Monodonta lineata (da Costa)
   Cantharidus striatus (L.)
   Cantharidus clelandi (Wood)
   Calliostoma zizyphinum (L.)
   Calliostoma papillosum (da Costa)

Family Turbinidae
   Tricolia pullus (L.)

Superfamily Neritacea

Family Neritidae
   Theodoxus fluviatilis (L.)

Order MONOTOCARDIA
The next twelve superfamilies are those referred to as mesogastropods.

Superfamily Architaenioglossa

Family Viviparidae
   Viviparus contectus (Müller)
   Viviparus viviparus (L.)

Superfamily Valvatacea

Family Valvatidae
   Valvata cristata Müller
   Valvata piscinalis (Müller)
   Valvata macrostoma Mörch

Superfamily Littorinacea

Family Lacunidae
   Lacuna pallidula (da Costa)
   Lacuna crassior (Montagu)
   Lacuna vincta (Montagu)
   Lacuna parva (da Costa)

Family Littorinidae
   Littorina littoralis (L.)
   Littorina saxatilis (Olivi)
   Littorina neritoides (L.)
   Littorina littorea (L.)

Family Pomatiasidae
   Pomatias elegans (Müller)

Family Aciculidae
   Acicula fusca (Montagu)

Superfamily Rissoacea

    Family Hydrobiidae
        Truncatella subcylindrica (L.)
        Hydrobia ulvae (Pennant)
        Hydrobia ventrosa (Montagu)
        Potamopyrgus jenkinsi (Smith)
        Pseudamnicola confusa (Frauenfeld)
        Bythinella scholtzi (Schmidt)

    Family Bithyniidae
        Bithynia leachi (Sheppard)
        Bithynia tentaculata (L.)

    Family Rissoidae
        Alvania crassa (Kanmacher)
        Alvania punctura (Montagu)
        Rissoa parva (da Costa)
        Rissoa inconspicua Alder
        Rissoa guerini Récluz
        Rissoa membranacea (Adams)
        Rissoa lilacina Récluz
        Cingula cingillus (Montagu)
        Cingula semicostata (Montagu)
        Cingula alderi (Jeffreys)
        Cingula semistriata (Montagu)

    Family Barleeidae
        Barleeia rubra Adams

    Family Assimineidae
        Assiminea grayana Fleming

The next four families are only provisionally placed in the Superfamily Rissoacea.

    Family Cingulopsidae
        Cingulopsis fulgida (Adams)

    Family Skeneopsidae
        Skeneopsis planorbis (Fabricius)

    Family Omalogyridae
        Omalogyra atomus (Philippi)
        Ammonicera rota (Forbes & Hanley)

    Family Rissoellidae
        Rissoella diaphana (Alder)
        Rissoella opalina (Jeffreys)

Superfamily Cerithiacea
   Family Turritellidae
      Turritella communis Risso

   Family Caecidae
      Caecum glabrum (Montagu)
      Caecum imperforatum (Kanmacher)

   Family Cerithiidae
      Bittium reticulatum (da Costa)

   Family Cerithiopsidae
      Cerithiopsis tubercularis (Montagu)

   Family Triphoridae
      Triphora perversa (L.)

Superfamily Strombacea
   Family Aporrhaidae
      Aporrhais pespelicani (L.)
      Aporrhais serresiana (Michaud)

Superfamily Calyptraeacae
   Family Trichotropidae
      Trichotropis borealis Broderip & Sowerby

   Family Capulidae
      Capulus ungaricus (L.)

   Family Calpytraeidae
      Calyptraea chinensis (L.)
      Crepidula fornicata (L.)

Superfamily Lamellariacea
   Family Lamellariidae
      Velutina velutina (Müller)
      Velutina plicatilis (Müller)
      Lamellaria perspicua (L.)
      Lamellaria latens (Müller)

   Family Eratoidae
      Erato voluta (Montagu)
      Trivia monacha (da Costa)
      Trivia arctica (Montagu)

Superfamily Cypraeacea
   Family Cypraeidae
      Simnia patula (Pennant)

Superfamily Naticacea
    Family Naticidae
        Natica alderi Forbes
        Natica catena (da Costa)
        Natica montagui Forbes
        Natica fusca Blainville
        Natica pallida Broderip & Sowerby

Superfamily Ptenoglossa
    Family Scalidae
        Clathrus clathrus (L.)

    Family Ianthinidae
        Ianthina janthina (L.)
        Ianthina exigua Lamarck
        Ianthina pallida Thompson

Superfamily Aglossa
    Family Eulimidae
        Eulima trifasciata (Adams)
        Balcis alba (da Costa)
        Balcis devians (Monterosato)

    Family Stiliferidae
        Pelseneeria stylifera (Turton)

The next three superfamilies are those included as neogastropods.

Superfamily Muricacea
    Family Muricidae
        Trophon muricatus (Montagu)
        Nucella lapillus (L.)
        Urosalpinx cinerea (Say)
        Ocenebra erinacea (L.)

Superfamily Buccinacea
    Family Buccinidae
        Buccinum undatum L.
        Chauvetia brunnea (Donovan)
        Neptunea antiqua (L.)
        Colus gracilis (da Costa)
        Colus jeffreysianus (Fischer)

    Family Nassidae
        Nassarius reticulatus (L.)
        Nassarius incrassatus (Ström)
        Nassarius pygmaeus (Lamarck)

Superfamily Toxoglossa
Family Conidae
Lora turricula (Montagu)
Philbertia teres (Forbes)
Philbertia gracilis (Montagu)
Philbertia linearis (Montagu)
Mangelia nebula (Montagu)
Mangelia powisiana (Dautzenberg)
Mangelia attenuata (Montagu)
Mangelia costulata (Risso)
Mangelia coarctata (Forbes)
Mangelia brachystoma (Philippi)

Subclass OPISTHOBRANCHIA
Order TECTIBRANCHIA
Superfamily Cephalaspidea
Family Acteonidae
Acteon tornatilis (L.)

Family Pyramidellidae
Turbonilla elegantissima (Montagu)
Chrysallida spiralis (Montagu)
Chrysallida obtusa (Brown)
Chrysallida indistincta (Montagu)
Chrysallida decussata (Montagu)
Odostomia unidentata (Montagu)
Odostomia lukisi Jeffreys
Odostomia turrita Hanley
Odostomia plicata (Montagu)
Odostomia eulimoides Hanley
Odostomia scalaris Macgillivray

Key to Families—A, Marine

1. Animal with clearly visible shell, perhaps temporarily covered by mantle folds when creeping; these withdraw to reveal shell when touched .   **2**

   Animal with no visible shell, looking like a dorid, though an internal shell may be felt through mantle, which is permanently fused over it
   Lamellariidae (p. 82)

2. Shell a short curved tube, open at one end, plugged with conical cap at the other .   .   .   .   Caecidae (p. 76)

   Shell not of this shape **3**

3. Shell with marginal slit, or with one or several holes in addition to mouth   **4**

   Shell without either slit or hole apart from mouth **7**

4. Shell with marginal slit **5**

   Shell with one or more holes **6**

5. Shell wound in a spiral; animal with operculum (Fig. 11)
   Scissurellidae (p. 36)

   Shell cap or limpet-shaped; animal without operculum Fissurellidae (p. 37)

6. Shell with one hole .   .   .   .   Fissurellidae (p. 37)

   Shell with more than one hole .   .   .   .   Haliotidae (p. 35)

7. Shell showing spire of several turns; animals with or without an operculum (usually with) **8**

   Shell without a spire, though apex may show a slight spiral twist   .   **9**

8. Shell polished; body whorl very large in relation to spire; mouth long and narrow; animal without operculum; mantle extends to cover shell in active animal .   .   .   .   Eratoidae (p. 84)

   Spire prominent; operculum usually present; mantle does not extend over shell or only slightly **15**

9. Shell ovoid or spindle-shaped with long, narrow mouth; no spiral structure visible; animal without operculum; mantle extends to cover shell in active animal **10**

   Shell conical or cap-shaped, no spire, sometimes with apex showing a slight spiral twist; no operculum **11**

   c

10. Shell ovoid, rounded at each end; with transverse ridges   Eratoidae (p. 84)

Shell elongated, drawn out at each end; smooth   .   . Cypraeidae (p. 85)

11. Shell conical (Fig. 13), apex not always central and may be tilted backwards; internal part of shell not divided by internal partition **12**

Shell cap-shaped (Figs 70, 74), apex often showing slight spiral coiling; internal part of shell may or may not be divided by internal partition (which may be felt under posterior end of foot) **14**

12. Animal with ctenidium in mantle cavity dorsal to head; no gills on edge of mantle skirt; shell depressed, without prominent radiating ribs
Acmaeidae (p. 38)

Animal without ctenidium in mantle cavity; shell not markedly depressed and may even be much raised **13**

13. Animal with accessory gills on mantle skirt; large animals, 10–25 mm or more when mature, littoral   .   .   .   . Patellidae (p. 40)

Animal without accessory gills on mantle skirt; small animals, not more than 10 mm when mature, not littoral   .   .   .   .   . Lepetidae (p. 43)

14. Shell with internal partition (which may be felt under posterior end of foot), not covered with periostracum; snout of animal normal, neck expanded into lateral lobes   .   .   .   . Calyptraeidae (p. 80)

Shell without internal partition, covered with periostracum; snout elongated into grooved proboscis, neck normal   .   .   .   .   . Capulidae (p. 79)

15. Shell without siphonal canal **16**

Shell with siphonal canal or notch **37**

16. Animal temporarily covers shell with mantle folds; shell ear-shaped, mouth very large and spire very short; shell covered with periostracum; mantle edges thick, swollen   .   .   .   . Lamellariidae (p. 82)

Shell not, or only slightly, covered by mantle or pedal lobes, mantle edges not thick **17**

17. Animal with at least 3 pairs of epipodial tentacles; neck lobe present behind each cephalic tentacle; operculum either a calcareous knob or, if horny, with at least 10 turns **18**

Animal without epipodial tentacles and neck lobes; operculum horny, with less than 10 turns, or absent **19**

18. Animal with horny operculum   .   .   .   . Trochidae (p. 44)

Animal with calcareous operculum   .   .   .   .   . Turbinidae (p. 52)

19. Shell with tall spire, at least 10 whorls, giving awl-shape; sutures very
shallow **20**

Shell with short spire, less than 10 whorls, giving squatter shape; whorls
often dip to sutures **23**

20. Shell smooth, without ribs or tubercles . . . . . Eulimidae (p. 90)

Shell with ribs, or spiral ridges, or both, giving tubercles **21**

21. Shell with ribs, either slight or prominent, never crossed by spiral lines;
white or pale brown in colour; mantle edge plain **22**

Shell with ribs crossed by spiral ridges on upper parts of whorls; with spiral
ridges only on basal part of body whorl; reddish-brown in colour; mantle
edge plain; opercular edge plain . . . . Cerithiidae (p. 77)

Shell with spiral ridges only; mantle edge bears pinnately branched tentacles;
lobe under right cephalic tentacle; opercular edge with fringe
Turritellidae (p. 74)

22. Ribs narrow and prominent; whorls tumid; mouth of shell circular; head
without mentum: large animals . . . . Scalidae (p. 87)

Ribs slight and slightly sinous; whorls nearly flat-sided; mouth of shell
ear-shaped; head with mentum; small animals . Pyramidellidae (p. 105)

23. Animal with mentum; eyes lie between tentacles; shell usually with small
tooth on columella . . . . Pyramidellidae (p. 105)

Shell and animal not like this **24**

24. Shell purple and white, fragile; animal pelagic with mucous float, cast
ashore on SW beaches only . . . . . . Ianthinidæ (p. 88)

Shell and animal not like this **25**

25. Shell covered with brown periostracum drawn out into bristles set in spiral
rows round whorls; animal with non-retractile grooved proboscis curving
from mouth to right side; not littoral . . . . Trichotropidae (p. 78)

Shell and animal not like this **26**

26. Body whorl equals about seven-eighths of shell height; shell partly covered
with folds when animal creeps; mantle cavity opens anteriorly, not to left;
tentacles flat lamellae; no umbilicus . . . . Acteonidae (p. 103)

Shell and animal not like this **27**

27. Initial 2–3 whorls of shell from narrow style-like summit to spire; animal
parasitic on echinoids . . . . Stiliferidae (p. 91)

Shell and animal not like this **28**

28. Shell smooth, umbilicate; animal with greatly enlarged propodium forming a bulldozer-like shield over the whole head except for the tips of the tentacles; on sandy beaches or dredged from sandy bottoms

Naticidae (p. 86)

Shell and animal not like this **29**

29. Shell with elevated spire **31**

Shell a plane spiral or nearly so; minute, about 1 mm **30**

30. Shell a plane spiral . . . . Omalogyridae (p. 72)

Shell with a very low spire . . . . Skeneopsidae (p. 72)

31. Mantle edge without a tentacle on either right or left side **33**

Mantle edge with a small tentacle on the right side, sometimes with a similar one on the left; often with a metapodial tentacle at hinder end of foot; small animals, not more than 6 mm **32**

32. Metapodial tentacle absent; in estuarine habitats   Hydrobiidae (p. 61)

Metapodial tentacle present, though it may be short; not estuarine

Rissoidae (p. 65)

33. Two metapodial tentacles present; shell with prominent curved umbilicus

Lacunidae (p. 56)

No metapodial tentacles **34**

34. Animal with eyes at tip of tentacles; groove from mantle cavity to sole of foot on each side of body; lives high up on beach   Assimineidae (p. 71)

Animal with eyes at base of tentacles; no grooves **35**

35. Shell minute (about 1 mm) with brown bands on light background; operculum pale; prominent glandular pore in middle of pedal sole and groove thence to posterior tip; male aphillic . . Cingulopsidae (p. 71)

Shell not so small (about 3 mm at smallest); foot without pore; penis in male (though it may be reduced outside breeding season) **36**

36. Shell small (about 3 mm), usually dark red, about 5 whorls; operculum concentric, deep crimson; opercular lobes of foot also dark; posterior end of foot slightly bifid . . . . Barleeidae (p. 71)

Shell not so small (minimum about 4 mm when mature); operculum spiral, not crimson; posterior end of foot not bifid . . . Littorinidae (p. 57)

7. Shell and animal dextral **38**

Shell and animal sinistral . . . .   Triphoridae (p. 77)

8. Shell with outer lip splayed out to form wide, palmate extension
Aporrhaidae (p. 78)

Shell with mouth not like this **39**

9. Shell with about 10 whorls; mouth with notch only for accommodation of siphon . . . .   Cerithiopsidae (p. 77)

Shell with 7 whorls at most; siphonal canal present **40**

0. Animal without operculum. . . . .   Conidae (p. 99)

Animal with operculum. **41**

1. Foot with 2 metapodial tentacles . . . .   Nassidae (p. 98)

Foot without metapodial tentacles **42**

2. Shell covered with yellow periostracal layer (commonly absent from triangular area on body whorl near inner lip); spiral lines form the only sculpture; outer lip thin . . . . . .   Buccinidae (*Colus*, p. 95)

Shell devoid of periostracum **43**

3. Shell with spiral keel round base **44**

Shell without such a keel **45**

4. Shell small (about 30 mm), with spiral ridges and growth lines but no ribs; outer lip thick when mature with tubercles on its inside
Muricidae (*Nucella*, p. 93)

Shell large (up to 75 mm), with spiral ridges and ribs in adapical part of body whorl and all parts of older whorls: outer lip without tubercles
Buccinidae (*Buccinum*, p. 95)

5. Shell with marked rib on outer lip . .   Buccinidae (*Chauvetia*, p. 95)

Shell without labial rib **46**

6. Shell with long siphonal canal ($= \frac{1}{4}$ of shell height), open or closed
Muricidae (p. 93)

Shell with siphonal canal much less than one quarter of shell height, always open **47**

47. Shell small (not more than 25 mm), with well marked ribs and spiral striae **48**

Shell large (up to 75 mm), with slight spiral striae and lines of growth but no ribs: apex slightly styliform; siphonal canal short with angulated junction to body whorl . . . . . . Buccinidae (*Neptunea*, p. 95)

48. Shell turreted; rather slender; mouth (with siphonal canal) occupies less than half total height of shell; small notch where outer lip meets body whorl; cream in colour . . . . Conidae (*Lora*, p. 99)

Shell not turreted, rather broad, whorls convex; mouth (with canal) occupies more than half total height of shell; no notch in outer lip; ashen grey in colour . . . . Muricidae (*Urosalpinx*, p. 93)

Key to Families—B, Freshwater

1. Shell with D-shaped mouth (Fig. 25), the columellar area expanded to form plate-like projection extending half-way across mouth; spire low; shell with speckled colour pattern . . . . Neritidae (p. 52)

   Shell with round or oval mouth, columellar region not expanded; colour pattern may be plain or banded, but not speckled 2

2. Shell appreciably taller than broad; gill does not extend beyond edge of shell in creeping animal 3

   Shell broader than tall, or breadth and height about equal; gill and a pallial tentacle (on right) extend beyond edge of shell as animal creeps
   Valvatidae (p. 55)

3. Operculum with concentric growth lines 4

   Operculum with spiral line 5

4. Shell large (30 mm or more), banded; operculum horny   Viviparidae (p. 54)

   Shell small (never more than 15 mm), unbanded; operculum partly calcified
   Bithyniidae (p. 64)

5. Eyes at tip of tentacles; ciliated groove runs from mantle cavity down each side of foot to sole . . . . Assimineidae (p. 71)

   Eyes at base of tentacles; no ciliated grooves on side of foot
   Hydrobiidae (p. 61)

Key to Families—C, Terrestrial

1. Mouth round, whorls convex, dipping to sutures; found in hedgerows on calcareous soils and chalk formations . . . . Pomatiasidae (p. 60)

    Mouth elongated, whorls flat; found amongst wet decaying leaves and wet moss on more acid soils . . . . Aciculidae (p. 60)

Systematic Part

HALIOTIDAE

Shell ear-shaped, with rapidly expanding whorls, mother-of-pearl sheen internally, with a row of small holes. No operculum. Shell muscle paired, right much larger than left. Mantle skirt split under hole. Pallial organs double, though those on the left are larger. Mantle edge thickened. Epipodium and epipodial tentacles well developed. Radula rhipidoglossan. Oesophageal gland large and papillate. Rectum runs through ventricle. Nervous system very unconcentrated with scalariform arrangement in foot. Larva a trochophore.

The only animal in this family which may be included here is the ormer, *Haliotis tuberculata* L., which is found only in the Channel Islands and does not occur on the British or Irish mainland. The shell shows a spiral structure, but expands rapidly; it bears a series of holes under one of which lies the exhalant opening of the mantle cavity; older holes become sealed with shelly material. The foot bears epipodial tentacles. Ormers browse on seaweeds, especially more delicate red weeds. Found at LWST and below on rocky shores. Animal whitish with green or reddish flecks. Up to 100 mm.

SCISSURELLIDAE

Shell spirally wound with 2–5 whorls, its mouth with slit at end of slitband. Operculum horny. Two shell muscles. Mantle skirt split under shell slit. Pallial organs double, gill feathery. Epipodial tentacles on foot. Radula rhipidoglossan. Rectum runs through ventricle.

There is only one British representative of this family, *Scissurella crispata* Fleming (Fig. 11), found not uncommonly in dredgings from muddy bottoms in the Western Approaches and off north and west coasts, and recognizable immediately by the small, colourless, spiral shell with marginal slit and prominent slitband on older whorls. 2 mm.

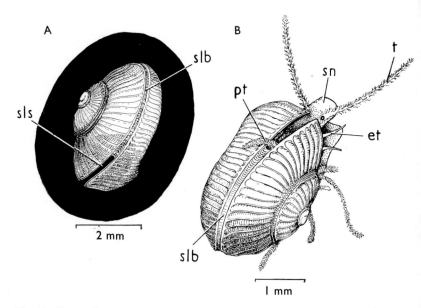

Fig. 11. *Scissurella crispata.* **A,** shell; **B,** living animal. *et,* epipodial tentacle; *pt,* palial tentacle; *slb,* slit band; *sls,* slit in shell; *sn,* snout; *t,* cephalic tentacle. From *British Prosobranch Molluscs* by Fretter and Graham.

FISSURELLIDAE

Shell conical or cap-shaped with apical or subapical hole or marginal slit at end of slitband; no operculum; shell muscle single, horseshoe-shaped. Mantle skirt split under slit or hole. Pallial organs double. Animal symmetrical. Mantle edge thickened. Foot with epipodial tentacles. Radula rhipidoglossan. Nervous system in foot scalariform. Rectum runs through ventricle. Left kidney much reduced and functionless. Larva a trochophore.

1. Shell with slit . . . . *Emarginula* sp.

   Shell with one hole **2**

2. Shell with backwardly directed apex; 1 hole on anterior slope
   Puncturella noachina

   Shell conical, with 1 apical hole. . . . . *Diodora apertura*

The family includes limpet-like animals adapted for a rock-clinging life; restricted to low levels on the shore (or living sublittorally) because of the hole in the shell, which acts as exhalant aperture. They feed on sponges.

Emarginula reticulata Sowerby

Shell with apex tilted backwards, about 30 radiating ribs; slit anteriorly over mantle cavity. A row of epipodial tentacles on each side of the foot. Animal white. Lives on underside of pitted rocks at LWST and below, often where there is a fine deposit of silt. On all rocky coasts. 10 mm long.

Emarginula conica and *E. crassa*

These are much rarer species of the genus, to be collected only by dredging.

Puncturella noachina (L.)

Shell cap-shaped with apex tilted backwards, about 30 radiating ribs; hole on anterior slope, drawn out to point anteriorly. A row of epipodial tentacles on each side of foot. Animal white-cream. Little known about mode of life. Not common; dredged on hard bottoms off Scottish coasts. 8 mm long.

Diodora apertura (Montagu)—Keyhole limpet

Shell conical, slightly broader posteriorly, and with oval mouth, with numerous ribs radiating from apical hole which is slightly figure-of-eight shaped. A row of epipodial tentacles on each side of foot. Mantle edge below shell greatly thickened and warty. Animal yellowish, often with brown speckles. Habitat and habits as for *Emarginula reticulata*. On all rocky coasts. Up to 30 mm long.

ACMAEIDAE

Shell conical or cap-shaped; no operculum. Shell muscle horseshoe-shaped. Mantle cavity small, over head, containing one ctenidium. No pallial gills. No epipodium. Radula docoglossan without rachidian tooth. Rectum does not run through ventricle. Larva a trochophore.

1. Shell with pink or brown rays on a pinkish background; inside white or pinkish; lips round mouth smooth, edge of mantle skirt smooth
Acmaea virginea

Shell with dark brown rays on white or greenish background: inside with chocolate brown mark; lips round mouth fringed, edge of mantle skirt with short tentacles; not found south of Humber or N. Wales   *A. tessulata*

The Acmaeidae are a group of animals with the same general mode of life as the common limpets but are not so resistant to desiccation, perhaps because they retain ctenidial respiration, and so live at lower levels on shores. Both species browse on diatoms and detritus.

Acmaea virginea (Müller) (Fig. 12)

Found on smooth stones at LWST and to a few metres depth on all coasts. 8 mm long.

A. tessulata (Müller)

Found on smooth stones at LWST and to a few metres depth. Northern. 8 mm long.

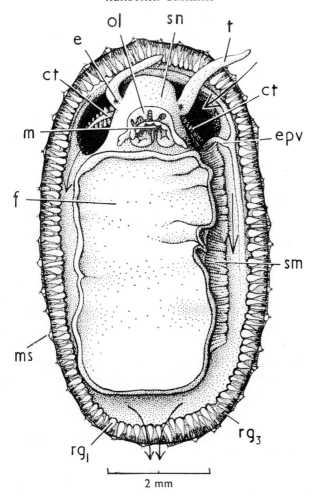

Fig. 12. *Acmaea virginea*. Ventral view. *ct*, ctenidium in mantle cavity; *e*, eye; *epv*, vein from mantle skirt to heart; *f*, foot; *m*, mouth; *ms*, cilia on sensory processes on mantle skirt; *ol*, outer lip; *rg₁*, *rg₃*, repugnatorial glands; *sm*, shell muscle; *sn*, snout; *t*, cephalic tentacle. From *British Prosobranch Molluscs* by Fretter and Graham.

PATELLIDAE

Shell conical or cap-shaped; no operculum. Shell muscle horseshoe-shaped. Mantle cavity small, over head, devoid of ctenidium. Pallial gills on edge of mantle skirt. No epipodium. Radula docoglossan with rachidian tooth. Rectum does not run through ventricle. Larva a trochophore.

1. Animal with pallial gills extending round whole extent of mantle skirt (Fig. 13); shell usually with radial ribs **2**

    Animal with pallial gills absent from mantle skirt anterior to head; shell smooth **4**

2. Pallial tentacles with white pigment **3**

    Pallial tentacles without white pigment; foot olive grey; inside of shell usually with grey green tint, head scar silvery; often living on high, dry rocks . . . . *Patella vulgata* (p. 42)

3. Pallial tentacles pale white; foot cream-apricot in colour, inner surface of shell white, head scar cream; confined to lower part of beach, and fond of wet places, not between Isle of Wight and Humber   *P. aspera* (p. 42)

    Pallial tentacles opaque white; foot dark, inner surface of shell and head scar dark; edge of shell marked with brightly coloured rays; S.W. England, Wales and W. Ireland only. . . . . *P. intermedia* (p. 42)

4. Shell translucent, depressed, with bright blue rays; from fronds of *Laminaria* (occasionally *Fucus*) . . . . . . *Patina pellucida pellucida* (p. 42)

    Shell opaque, conical, without blue rays except at apex; from underneath holdfast of *Laminaria* . . . . *P. p. laevis* (p. 42)

Note that, whilst the different kinds of limpets in the genus *Patella* may usually be easily identified, there is a region of coast centring on the Isle of Wight where many intermediate forms abound and identification may not be easy or even possible.

The Patellidae constitute an outstandingly successful group of animals which have become adapted for life in intertidal conditions; some species can live high on the shore and withstand brackish water. They browse on weeds and on the film of algae and detritus covering the rocks on which they live. Each animal has its own home from which it makes feeding excursions, passing the intertidal period firmly clamped to a particular area of rock that its shell has grown to fit. The genus *Patella* contains rock limpets, *Patina* weed limpets.

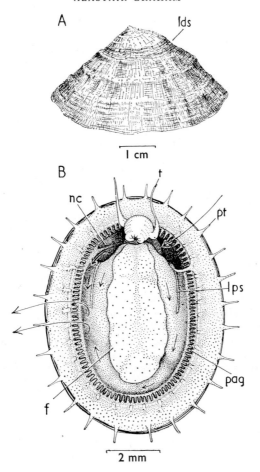

Fig. 13. *Patella vulgata.* **A,** shell in lateral view; **B,** living animal, ventral view; arrows show direction of ciliary currents. *f,* foot; *lds,* ledge due to uneven growth; *lps,* lateral pallial streak (in young animals only); *nc,* nuchal cavity; *pag,* pallial gill; *pt,* pallial tentacle; *t,* cephalic tentacle. From *British Prosobranch Molluscs* by Fretter and Graham.

Patella vulgata L.—Common limpet (Fig. 13)

Shell distinguished by the grey-green internal colour. The only species withou white pigment on the tentacles and the only one capable of surviving on expose rocks high on the beach. Those in this situation tend to have tall shells, those a lower levels are less tall. Breeds October to December. On rocks and stones an in pools from HWNT downwards. Can resist salinities $\lessdot$ 3‰. On all suitabl coasts. 30 mm long.

P. aspera Lamarck—China limpet

Shell distinguished by white internal surface often with bluish iridescence Confined to the lower part of rocky shores and prefers runnels of water to dr areas; restricted to pools at higher levels; avoids brackish water and extrem shelter. Breeds summer. On all suitable coasts from Isle of Wight west an north to Humber. 30 mm long.

P. intermedia Jeffreys—Black-footed limpet

Shell distinguished by bright marginal rays; animal by whiteness of pallia tentacles and dark pigmentation of foot. Widely spread over exposed rock shores in S.W. England and Wales, not occurring elsewhere. Recently recorde from Co. Mayo. Breeds throughout year with summer maximum. 25 mm long.

Patina pellucida (L.)—Blue-rayed limpet

This species is almost confined to *Laminaria* plants and occurs in two forms *P. p. pellucida* from the surface of the fronds, and *P. p. laevis* from cavities in th holdfast under the base of the stipe; *pellucida* has a smooth, horn-coloured shel with beautiful kingfisher-blue rays; *laevis* has a slightly rougher pale shell withou noticeable blue rays except at the apex. On all shores. 10 mm long.

LEPETIDAE

Shell small, cap-shaped. No operculum. Shell muscle horseshoe-shaped. Mantle cavity small, over head. Neither ctenidium nor pallial gills. No epipodium. Radula shows fusion of lateral teeth in each row. No eyes.

1. Shell depressed; inside simple **2**
   Shell conical with apex turned back: small internal partition at apex
Propilidium exiguum.

2. Shell white . . . . *Lepeta caeca*
   Shell orange . . . . *L. fulva*

A small family of blind limpet-like animals only to be encountered in dredgings from off Scottish or Irish coasts, though *Propilidium exiguum* has been recorded from LWEST. All are rare.

D

TROCHIDAE

Shell spirally wound, often pyramidal, showing mother-of-pearl colouring internally, often revealed externally by wear of outer layer. Operculum poly-gyrous, horny. Mantle cavity with only 1 gill, the anterior part of which is bipectinate, the base monopectinate. Left kidney a papillary sac. Radula rhipidoglossan. Oesophageal gland large and papillate. Stomach with well-developed spiral caecum. Rectum passes through ventricle. Larva a trochophore or suppressed.

1. Shell with umbilicus or at least chink at base 2
   No umbilicus at base 4

2. Six pairs of epipodial tentacles (Fig. 5); shell small and pearly; left neck lobe with smooth edge . . . . *Margarites* (**a1**)

   Three pairs of epipodial tentacles, outside of shell not pearly; left neck lobe with fringed (Fig. 22) or scalloped (Fig. 4) edge 3

3. Spire rather low, umbilicus plain, shell with reddish stripes, no tooth on columella (Fig. 16) . . . . *Gibbula* (**b1**)

   Spire higher, umbilicus a chink, shell dark usually exposing mother of pearl at apex; tooth on columella (Fig. 15) . . *Monodonta lineata* (p. 48)

4. Three pairs of epipodial tentacles; oral lips without tubular extension
                                                     Cantharidus (**c1**)

   Four pairs of epipodial tentacles; oral lips drawn out into mid-ventral tubular extension bent to right side (Fig. 22) . . . *Calliostoma* (**d1**)

a1. Shell with smooth surface, polished, translucent; animal orange in colour
                                                  Margarites helicinus (p. 46)
   Shell showing numerous spiral ridges, opaque; animal cream in colour
                                                       M. groenlandicus

b1. Shell turreted, with tubercles on upper part of each whorl **b2**
   Shell without tubercles, not turreted **b3**

b2. Large shell (up to 20 mm) with prominent spiral keel marking periphery of body whorl, tubercles prominent, mouth oblique because of advancement of outer lip over inner; usually dredged but may be littoral; left neck lobe fringed . . . . *Gibbula magus* (p. 46)

   Small shell (up to 8 mm) with slight peripheral keel, tubercles not prominent, mouth not oblique; always dredged; left neck lobe scalloped
                                                           G. tumida (p. 46)

b3. Shell with a few broad reddish lines running across each whorl; inner lip almost straight in basal view of shell (Fig. 17B) . *G. umbilicalis* (p. 46)

Shell with many narrow reddish lines running across each whorl; inner lip markedly curved in basal view of shell (Fig. 17A) . . *G. cineraria* (p. 46)

Shell like *G. umbilicalis*; streaks coloured purple; umbilicus nearly closed; found in Channel Islands only . . . . . . *G. pennanti* (p. 46)

c1. Shell rather narrow at base (greatest ht = about 2x greatest breadth), 8–9 spiral ridges on body whorl, the most basal slightly larger than others and making a keel; dark brown bands of colour cross these ridges
Cantharidus striatus (p. 48)

Shell broad at base (greatest ht = greatest breadth), 6–7 spiral ridges on body whorl, the most basal distinctly larger than others and making a marked keel; spots, speckles or short bands of red-purple cross the ridges . . . . *C. clelandi* (p. 48)

d1. Shell with smooth spiral ridges; regularly conical; tip of snout finely fringed, littoral or dredged . . . . *Calliostoma zizyphinum* (p. 50)

Shell with spiral ridges beset with small papillae; sides of cone slightly incurved (coeloconic); tip of snout with many long finger-shaped tentacles; never littoral . . . . *C. papillosum* (p. 50)

The family Trochidae contains the shells known as top shells because of their resemblance to spinning tops. The animals in this family, though classified as Diotocardia are almost at the monotocardian grade of organization, having lost the right ctenidium and osphradium, though they retain the right auricle. The left ctenidium, too, has become modified in that only its free tip retains the primitive bipectinate condition. The animals are vegetarians and feed by collecting small algal cells and vegetable detritus by means of their rhipidoglossan radula.

Margarites helicinus (Fabricius) (Figs 5, 14)

Shell small, globular with green-violet iridescence, whorls rounded and without surface markings. Gregarious under stones or amongst weed. Locally abundant north of Humber and Bristol Channel, at LW to a few metres. 4 mm.

M. groenlandicus (Gmelin)

Shell small, with low spire, whorls marked plainly with spiral ridges, with reddish tint. Under stones and in *Laminaria* holdfasts at LWEST and below. Rare. 5 mm.

Gibbula magus (L.) (Fig. 19)

Shell slightly turreted, with marked keel round periphery of body whorl and pronounced tubercles on apical part of each whorl. Grey with reddish stripes. Left neck lobe fringed, right plain. Confined to south-west and west coasts and usually dredged from muddy sand; in some areas occurs littorally at LWEST. 20 mm.

G. tumida (Montagu)

Shell turreted and with keel round base of body whorl; sometimes with small tubercles on apical part of each whorl. Grey with chocolate brown stripes. Left neck lobe scalloped, right plain. The animal may be dredged from gravelly or shelly bottoms, commoner in north than south. 8 mm.

G. umbilicalis (da Costa) (Figs 17B, 18)

Shell with whorls flattened so as to present a smooth curved side in profile; proportions of cone vary with habitat. Distinguished from next species by coarse red-purple markings and by straightness of inner lip of shell mouth when examined in a basal view. In upper parts of rocky beaches (HWN—downwards, less common below MLWN) local but often abundant. Eats algae and vegetable detritus. Absent from east coasts. 10 mm.

G. cineraria (L.) (Figs 16, 17A)

Shell shape as in last species but distinguished by many very narrow bands of colour on shell and by curvature of inner shell lip when seen in basal view. On lower part of shore (MLWN downwards). With same habits as *G. umbilicalis*. Found on all rocky shores. 10 mm.

G. pennanti (Philippi)

Shell like that of *G. umbilicalis* save that streaks are purple and the umbilicus is nearly closed. Found in Channel Islands only, not on mainland of Britain. 10 mm.

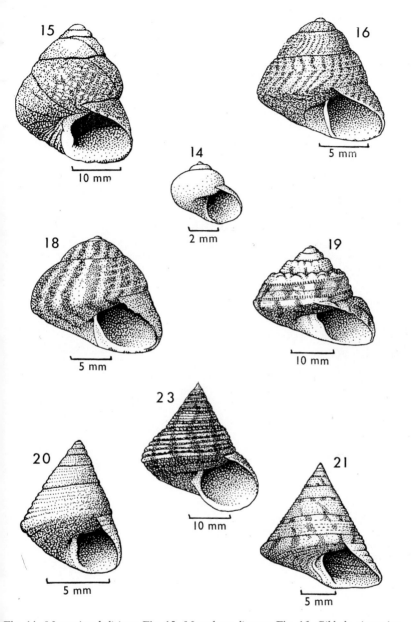

Fig. 14, *Margarites helicinus*; Fig. 15, *Monodonta lineata*; Fig. 16, *Gibbula cineraria*;
Fig. 18, *Gibbula umbilicalis*; Fig. 19, *Gibbula magus*; Fig. 20, *Cantharidus striatus*;
Fig. 21, *Calliostoma zizyphinum*; Fig. 23, *Calliostoma papillosum*.

Monodonta lineata (da Costa) (Fig. 15)

Shell with tall spire, whorls smooth except for growth marks, slightly swollen so as to dip slightly to sutures. Umbilicus slight. Columellar lip shows marked tooth at basal end. Dark, with numerous wavy streaks, base pale and often showing mother-of-pearl hues. Apex often also pearly. Occurs on rocks and stones, often on upper surfaces, about HWNT, between Poole and Mersey. Common where found, but markedly local, and badly affected by cold of 1963. 25 mm.

Cantharidus striatus (L.) (Fig. 20)

Shell narrowly conical, straight-sided, body whorl with sharp peripheral keel. Grey with few broad chocolate-brown vertical bands. Dredged from soft bottoms in S.W. England only. 10 mm.

C. clelandi (Wood) (Fig. 4)

Shell equiangular cone, straight-sided, body whorl with bulging peripheral keel. Light with short red streaks. Dredged on muddy gravel. Not found off south-east coasts, commoner in north than south. 10 mm.

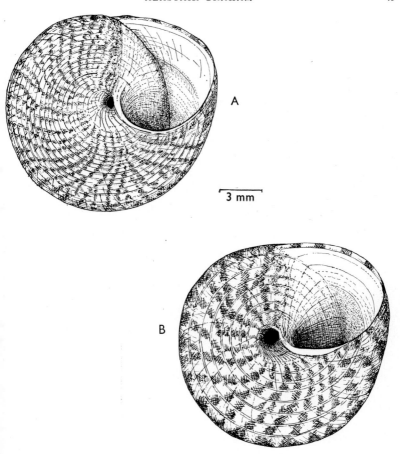

3 mm

Fig. 17. Shells of *Gibbula*: **A**, *G. cineraria*; **B**, *G. umbilicalis*. From *British Prosobranch Molluscs* by Fretter and Graham.

Calliostoma zizyphinum (L.) (Fig. 21)

Regularly conical shell, straight-sided, body whorl with prominent peripheral keel. No umbilicus. Colour light with short streaks which are red on keel and may be red or brown elsewhere. Occasional dark specimens are found without streaks, as well as white or purple ones. Tip of snout slightly fringed; ventral lip drawn out into spout-like projection to right. Under stones at low water of spring tides and below. Tolerates some brackishness. Common on all rocky shores. 25 mm.

C. papillosum (da Costa) (Figs 22, 23)

Regularly conical shell though often showing slight inward curvature of sides (coeloconoid). Body whorl keeled, but not prominently; other spiral ridges all marked by small rounded bosses or papillae. No umbilicus. Shell pinkish with darker bands. Tip of snout markedly fringed; ventral lip drawn out into spout-like projection to right. Dredged from sandy gravel, not littoral. West coasts only. 30 mm.

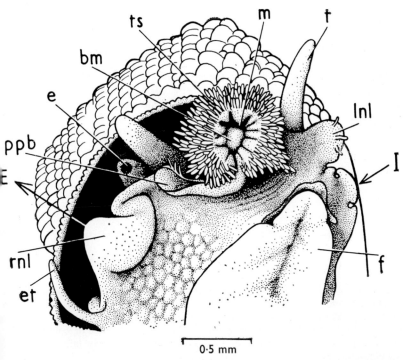

Fig. 22. *Calliostoma papillosum*, anterior part of body, ventral view. *bm*, anterior tip of buccal mass; *e*, eye; *E*, exhalant current from mantle cavity; *et*, epipodial tentacle; *f*, foot; *I*, current entering mantle cavity; *lnl*, left neck lobe; *m*, mouth; *ppb*, proboscis; *rnl*, right neck lobe; *t*, tentacle; *ts*, papillae on snout. From *British Prosobranch Molluscs* by Fretter and Graham.

TURBINIDAE

Shell spirally wound, turban-shaped or conical. Operculum calcareous. Soft parts almost exactly as in trochids.

Only one British turbinid exists, the small animal *Tricolia pullus* (L.) (Fig. 24) which may be recognized immediately by the rounded, white, calcareous operculum plugging the shell mouth. The shell has numerous irregular reddish brown streaks on a white background. The animal has 3 pairs of epipodial tentacles. Neck lobes both fringed. In rock pools or amongst weed, especially *Chondrus crispus*, at LWST and below. On rocky coasts in south and west, much commoner in south. 7 mm.

NERITIDAE

Shell usually low, without umbilicus; columellar region grows across mouth to form a kind of septum which partially occludes it, outer lip often thickened; operculum calcareous. Soft parts highly modified. Mantle cavity with gill bearing double row of leaflets on left; penis lies median to base of right tentacle. Rectum passes through ventricle. Two auricles in heart. Eggs laid in capsules strengthened by calcareous material; development direct.

The family contains only one British species and is predominantly character-istic of warmer countries. There is a general trend for neritids to move from marine habitats to estuarine, freshwater and terrestrial. Their internal anatomy is highly modified in ways not paralleled in other groups and, though technically diotocardian, many have attained a monotocardian grade in a fashion different from that achieved by other molluscs.

Theodoxus fluviatilis (L.) (Fig. 25)

The only British neritid; found in fresh and estuarine conditions where the salinity does not exceed 6‰. It is limited to fresh water with high calcium content. It lives in situations where it is protected from water currents and light so that it is found under stones and on sunken wood, tree bases. The semilunar mouth, low spire, speckled colour pattern of shell identify it easily. Common where the right conditions exist. 10 mm.

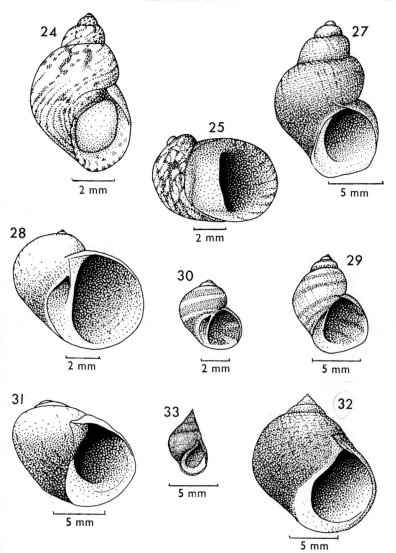

Fig. 24, *Tricolia pullus*; Fig. 25, *Theodoxus fluviatilis*; Fig. 27, *Lacuna crassior*; Fig. 28, *Lacuna pallidula*; Fig. 29, *Lacuna vincta*; Fig. 30, *Lacuna parva*; Fig. 31, *Littorina littoralis*; Fig. 32, *Littorina littorea*; Fig. 33, *Littorina neritoides*.

VIVIPARIDAE

Shell conical, with narrow umbilicus or none; whorls swollen. Operculum horny and concentric. Animal with long snout and tentacles; groove across floor of mantle cavity. Kidneys with long ureter. Right tentacle of male modified to form penis; female uses terminal part of oviduct as uterus and the young are born as miniature adults. Freshwater.

Sutures deep, umbilicus distinct, conical .   .   .   .   .   . *Viviparus contectus*

Sutures not deep, no umbilicus, ovoid .   .   .   . *V. viviparus*

Viviparus contectus (Millet)

The shell is conical, rather glossy, with very deep sutures and a distinct umbilicus; rather thin, with a sharp apex and a thin oval operculum; brown bands on shell not marked against dark blue-green background; body whorl (including aperture) about as broad as high. Occurs in hard water, with slight to moderate current, to the edge of brackish conditions, often burrowing slightly in gravel. Browses on detritus and is a facultative ciliary feeder. Not recorded further north than Lancashire and Yorkshire, nor in Wales and Ireland. 35 mm.

Viviparus viviparus (L.)

The shell is a more ovoid shape, not very glossy, with shallow sutures and no umbilicus; stout shell with rounded apex and a thick operculum not particularly oval in shape; brown bands on shell distinct against light greenish-yellow background; body whorl (including aperture) distinctly broader than high. Occurs in the same kind of habitat as the previous species, is more common and has a wider distribution though absent from Wales, Scotland and Ireland. 35 mm.

VALVATIDAE

Shell low with distinct umbilicus; operculum horny. Gill with double row of aflets; pallial tentacle present on right side of mantle skirt. No oesophageal and; stomach simple. Hermaphrodite. No free larval stages. Freshwater.

Shell flat, whorls slightly keeled, about 1 mm high . . .   *Valvata cristata*

Shell not flat, but with elevated spire; whorls not keeled; height more than 1 mm **2**

Shell about 6 mm high, surface not glossy; umbilicus narrow   *V. piscinalis*

Shell about 2 mm high, surface glossy; umbilicus widely open   *V. macrostoma*

Valvata cristata Müller

Shell planorboid, 5 whorls in one plane, slightly keeled at the periphery. Gill rotrudes as animal creeps. Moderately common in slowly running fresh water, 1 muddy environments of soft water, especially amongst the roots of flags, eeds. Shell light horn colour. Locally common except in Cornwall. 1 mm.

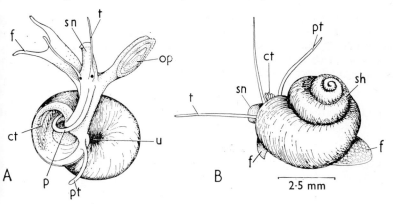

ig. 26. *Valvata piscinalis.* **A,** animal extended from shell; **B,** animal crawling. *ct*, anterior tip of ctenidium projecting from mantle cavity; *f*, foot; *op*, operculum; *p*, penis; *t*, pallial tentacle; *sh*, shell; *sn*, snout; *t*, cephalic tentacle; *u*, umbilicus. From *British Prosobranch Molluscs* by Fretter and Graham.

Valvata piscinalis (Müller) (Fig. 26)

Shell globular, 6 whorls, not keeled. Umbilicus is deep but has narrow mouth. Shell brownish yellow but nearly always covered with green coat of encrusting algae. Moderately common in soft waters, where there is some current, mainly on weeds, throughout Britain. 7 mm.

Valvata macrostoma Mörch

Recognizable from *cristata* by the slightly elevated spire, from *piscinalis* by the more depressed spire, the wide and deep umbilicus and its smaller size, not exceeding about 2 mm in height. In the sides and bottoms of ditches and drains, and on the leaves of plants. Prefers hard water; limited to S.E. England. Shell brown.

LACUNIDAE

Shell rather small, thin, smooth, often with umbilicus. Operculum horny, o few turns. Oesophageal gland present. Larva a veliger, though sometime suppressed.

1. Mouth of shell equal to whole shell height or nearly so   *Lacuna pallidul*

    Mouth of shell much less than this . . . .

2. Shell smooth, mouth out-turned, umbilicus obvious . . . . . .

    Shell with small vertical sinuous wrinkles rather obscured by periostracum umbilicus often absent, without brown bands . . . . . *L. crassio,*

3. Shell conical with 6 whorls, mouth rather large and angulated below umbilicus often with brown bands . . . . *L. vinct*

    Shell globular with 3–4 whorls, mouth very little out-turned or angulated below umbilicus, usually with brown bands . . . . . *L. parv*

A family of animals with habits very similar to those of littorinids thougl living at lower levels on the beach. The genus *Lacuna* is always recognizable b the two metapodial tentacles (Fig. 6). They lay jelly-like masses of spawn on th weeds on which they feed and are most common at spawning time, perhap retreating to deeper waters at other times.

Lacuna pallidula (da Costa) (Fig. 28)

Easily told from other species of *Lacuna* by the very large, out-turned mout which is equal to the total height of the shell. Lives in *Laminaria* zone. 8 mm.

L. crassior (Montagu) (Fig. 27)

Shell with both spiral and vertical markings, the latter the more prominent though obscured by periostracum which also shows puckering. Umbilicus may or may not be present. Sutures deep. Mouth slightly out-turned and angulated at base. The least common species and restricted to N. 8 mm.

L. vincta (Montagu) (Figs 6, 29)

Shell with distinctly conical shape, the whorls usually bearing brown spira bands. Mouth out-turned and angulated at base below umbilicus. The commones species, living on weeds, especially fucoids, *Ceramium* and *Polysiphonia*. 6 mm

L. parva (da Costa) (Fig. 30)

Shell with globular shape, the body whorl occupying almost all the height Sometimes with, sometimes without, spiral bands. Mouth not angulated below umbilicus. On weeds, especially *Chondrus*, at LWST on south and west coasts 5 mm.

LITTORINIDAE

Shell solid, of various shapes. Operculum horny with few turns. Radula very
ong. Oesophageal gland present. Larva a veliger, occasionally suppressed.

. Shell with obvious spire (Fig. 32), giving conical shape **2**

Shell globose, with spire so depressed that body whorl equals total shell
height (Fig. 31) .   .   .   .   *Littorina littoralis* (p. 58)

. Shell with tumid whorls which dip to the sutures (Fig. 34); outer lip curving
in to meet body whorl more or less at right angles; longitudinal black stripes
on tentacles.   .   .   .   .   *L. saxatilis* (p. 58)

Shell with flat-sided spire (Fig. 32), whorls hardly dipping to sutures; outer
lip straight, running apically to join body whorl at acute angle; tentacles
with either longitudinal or transverse black stripes **3**

. Shell with flat sides, dark columella, periostracum projecting to give free
flap round mouth; shiny black surface; animal with longitudinal black
stripe on tentacles .   .   .   .   *L. neritoides* (p. 58)

Shell with sides nearly flat but whorls show slight dipping to surface; surface
with spiral ridges, especially when young, columella white, no projecting
periostracum; animal with transverse black stripes on tentacles
L. littorea (p. 58)

The family Littorinidae contains the gastropods popularly called winkles,
constituting an extraordinarily successful group that abound on all beaches
except those that are sandy. They feed mainly on the film of detrital material on
rocks or mud surfaces, though they also rasp weeds. They have adapted to withstand desiccation and reduced salinity and hardly extend below LW mark.

Littorina littoralis (L.)—Flat winkle (Fig. 31)

Shell distinguished by extreme depression of spire; mouth out-turned; very variable in colour. Usually found on *Fucus vesiculosus* and *Ascophyllum nodosum* mimicking their air bladders; especially common where plants of these border rock pools. Abundant on all rocky or stony shores. 10 mm.

Littorina saxatilis (Olivi)—Rough winkle (Fig. 34)

Shell usually with spiral ridges and grooves; mouth out-turned especially at base; very variable in colour. Usually in cracks, crevices and empty barnacle shells high on rocky beaches in association with the weed *Pelvetia canaliculata* Viviparous, females being larger, with a more tumid body whorl than males Abundant on all rocky coasts except the most exposed. 8 mm.

Littorina neritoides (L.) (Fig. 33)

Shell smooth, covered by periostracum which projects at edge of mouth to give flexible flap; columella dark; spire conical, slightly coeloconoid. Found in rock crevices and cracks at and above HWST. Occurs on all rocky coasts, even exposed ones, rising higher in proportion to amount of exposure and consequent splash. Migrates to lower levels at breeding season (March–April). Up to 10 mm.

Littorina littorea (L.)—Edible winkle (Fig. 32)

The largest winkle. Shell recognizable by angle at which outer lip meets body whorl and by white columella. Spiral ridges which are marked in young animals tend to become obscured in older ones. Animal recognizable by transverse black barring of tentacles. Occurs on rocky, stony, muddy beaches from HWNT to LWST and below. Abundant everywhere. 25 mm.

The species of *Littorina* are usually easy to identify. Confusion may arise however when dealing with young specimens of *L. littorea* found at the same level as older specimens of *L. saxatilis*. At this size both are ridged. *L. littorea* can however be readily distinguished by the white columella and by the banding on the tentacles. There are numerous varieties of both *L. littoralis* and *L. saxatilis*; descriptions of the former will be found in Dautzenberg and Fischer (1914) and Barkman (1955), and of the latter in James (1968a, b).

It must be remembered that in spring specimens of *L. neritoides* may migrate to spawn and may be collected at levels usually more typical of *L. saxatilis* or even *L. littoralis*.

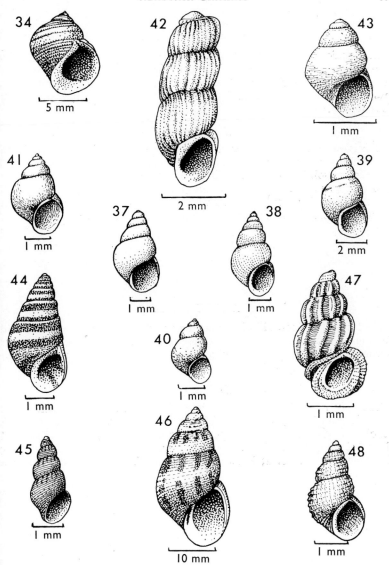

Fig. 34, *Littorina saxatilis*; Fig. 37, *Hydrobia ulvae*; Fig. 38, *Hydrobia ventrosa*; Fig. 39, *Potamopyrgus jenkinsi*; Fig. 40, *Bythinella scholtzi*; Fig. 41, *Pseudamnicola confuas*; Fig. 42, *Truncatella subcylindrica*; Fig. 43, *Cingula alderi*; Fig. 44, *Cingula cingillus*; Fig. 45, *Cingula semicostata*; Fig. 46, *Cingula semistriata*; Fig. 47, *Alvania crassa*; Fig. 48, *Alvania punctura*.

E

POMATIASIDAE

Shell conical, with marked spiral ridges; mouth circular; operculum spiral with calcareous layer. Ctenidium absent, though the osphradium persists; oesophagus without glands; crystalline style in stomach. Sole of foot divided into right and left halves moved alternately in walking. Egg with direct development.

Pomatias elegans (Müller)

This is one of only two species of terrestrial prosobranch found in Britain and has not been recorded from Scotland or Ireland. It is confined to situations with a calcareous soil (pH 7·5–7·9, calcium content over 6%), burrowing when cold or dry. It is recognizable by the operculum, the almost exactly circular mouth and the spiral ridges on the tumid whorls. Shell light brown with occasional darker streaks. Abundant where conditions are right. 15 mm.

ACICULIDAE

Shell small, cylindrical, with blunt apex; whorls flat-sided, marked with delicate lines of growth. Ctenidium absent, though the osphradium persists. Respiration largely by means of air bubble in mantle cavity. Oesophagus glandular, and no crystalline style. Direct development.

Acicula fusca (Montagu)

The only terrestrial prosobranch in addition to Pomatias elegans. Found amongst dead leaves, moss, especially where these are wet, in slightly acid conditions. It is recognizable by the operculum, the cylindrical shell with elongated mouth and its characteristic habitat. Shell brownish, glossy. Commoner in the north; known also from S.W. Scotland, Merioneth and Ireland. 2 mm.

HYDROBIIDAE

Shell conical, smooth. Operculum horny, spiral. Radula small. Oesophagus without gland; stomach with crystalline style. Nervous system concentrated. Penis simple. Usually fresh or brackish water animals with direct development.

1. Shell decollated (p. 59, Fig. 42), having lost apical whorls, remaining whorls with fine ribs . . . . . . *Truncatella subcylindrica* (p. 62)

Shell not decollated, with complete spire, smooth **2**

2. Mantle edge with tentacle on right side . . . . *Hydrobia* (**a1**)

Mantle edge without tentacle on right side **3**

3. Apex blunt, whorls swollen, dipping markedly to sutures; not taller than 3 mm; local to S. Lancs. and Stirlingshire . *Bythinella scholtzi* (p. 62)

Apex pointed, whorls not markedly swollen, often rather flat; taller than 3 mm **4**

4. Apical angle narrow, about 45°, sutures distinct but not deep, body whorl larger, sometimes with keel or bristles at periphery; snout dark, base of tentacles dark with light line near eye (Fig. 35A); in brackish or fresh water . . . . *Potamopyrgus jenkinsi* (p. 62)

Apical angle wide, about 60°, whorls high shouldered, the apical part of each flattening as it approaches suture; in slightly brackish or fresh water (recorded from East Anglia, S.E. Ireland only)
Pseudamnicola confusa (p. 62)

a1. Shell rather flat-sided; outer lip straight and approaches body whorl at acute angle; dorsal surface of head with lozenge-shaped dark mark between tentacles; tentacles with short length of black pigment near tip, left one thicker and more heavily ciliated than right (Fig. 35B); from wet mud flats where salinity still high though some freshwater influence exists
Hydrobia ulvae (p. 62)

Shell with swollen whorls dipping to sutures; outer lip curved and approaches body whorl nearly at right angles; snout with dark V-shaped mark between eyes bounded by light streaks extending back from base of tentacles; tentacles alike, without black marks (Fig. 35C); from mud and algal growths in lagoons where salinity is still high though some freshwater influence exists . . . . *H. ventrosa* (p. 62)

Truncatella subcylindrica (L.)—Looping snail (Fig. 42)

Shell of 3 whorls in spire with parallel sides; whorls with shallow longitudinal ribs; mouth small, ear-shaped. Young shells show a complete spire, but at maturity the apical whorls are broken off and the shell is then said to be decollated. Animals move like a looper caterpillar, using snout and foot. Snout very long and tentacles set far back. Operculum narrow, set across foot. A groove runs from the mantle cavity to the underside of the snout at the front of the foot. Confined to muddy places at high tide level where *Suaeda maritima* and *Halimione portulacoides* grow. South coast of England. 3 mm.

Hydrobia ulvae (Pennant) (Figs 35B, 36A, 37)

Can be distinguished from *H. ventrosa* by the lack of swelling of the whorls of the shell, by the straight outer lip where it meets the body whorl, by the pigmentation of snout and tentacles and by its occurrence on wet mud in positions which are usually estuarine but are sometimes marine. Occurs at about mid-tide level, crawling down beach as the tide ebbs, and being carried back up the beach by floating on the flowing tide. 4 mm.

Hydrobia ventrosa (Montagu) (Figs 35C, 38)

Distinguished from previous species by the tumid whorls, the curvature of the outer lip where it meets the body whorl, the pigmentation of snout and tentacles and the tendency to occur in lagoons without direct contact with clearly marine conditions. Abundant where it occurs, but distinctly local. 4 mm.

Potamopyrgus jenkinsi (Smith) (Figs 35A, 39)

Shell usually black because of deposits (which may obscure shape altogether) but yellowish when clean. Body whorl long, occupying nearly two-thirds of shell height. Convexity of whorls and the presence of a keel seem to be correlated with environment, those from brackish water tending to more tumid whorls and keels, those from fresh being more slender, unkeeled and paler in colour. The animals abound on stones or weed or in mud at the bottom of running water. All (except one specimen found in 1958) are females reproducing parthenogenetically and viviparously. 5 mm.

Pseudamnicola confusa (Frauenfeld) (Fig. 41)

The flattening of the whorls immediately below the suture is the main characteristic of this shell which is thin, glossy and translucent. The shell has an apical angle much greater than that of *Hydrobia* spp. or *Potamopyrgus* and the body whorl occupies about half the shell height. Found only in muddy ditches which are brackish or nearly totally fresh in East Anglia and S.E. Ireland. 3 mm.

Bythinella scholtzi (Schmidt) (Fig. 40)

Shell recognizable by blunt apex, small size and well marked sutures. The animal has tentacles which contract down to rounded knobs. Found only in canals in the Manchester district and near Grangemouth, Stirlingshire. Shell brown, but usually blackened by deposits. 2–3 mm.

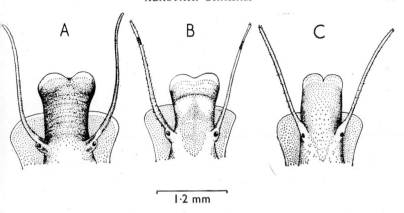

Fig. 35. Dorsal view of head of **A,** *Potamopyrgus jenkinsi*; **B,** *Hydrobia ulvae*; **C,** *Hydrobia ventrosa* to show the pattern of pigmentation and ciliation. From *British Prosobranch Molluscs* by Fretter and Graham.

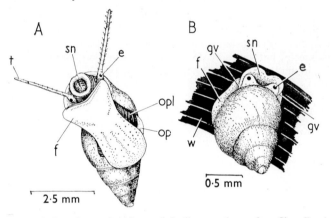

Fig. 36. **A,** *Hydrobia ulvae* swimming and feeding on the surface film; **B,** *Assiminea grayana*, crawling. *e*, eye; *f*, foot; *gv*, groove from mantle cavity to edge of foot; *op*, operculum; *opl*, opercular lobes of foot; *sn*, snout; *t*, cephalic tentacle; *w*, weed. From *British Prosobranch Molluscs* by Fretter and Graham.

BITHYNIIDAE

Shell conical, smooth. Operculum at least partly calcified, concentric. Oesophagus without gland; stomach with crystalline style. Nervous system concentrated. Penis branched. Freshwater animals with direct development.

Shell with well developed umbilicus, mouth rounded where outer lip approaches body whorl; whorls dip markedly to sutures; not more than 7 mm high . . . . *Bithynia leachi*

Shell with hardly any umbilicus, mouth narrow and pointed where outer lip approaches body whorl; whorls dip slightly to sutures; more than 7 mm high . . . . *B. tentaculata*

Bithynia leachi (Sheppard)

Recognizable by umbilicus and shape of mouth; moderately glossy, 5 whorls; found in slow rivers, canals, ponds with hard water; on vegetation in summer, overwintering in mud. Rather local in S. and E. England. 6 mm.

Bithynia tentaculata (L.)

Shell glossy, 6 whorls, without more than a small chink to represent the umbilicus, and with angulated mouth. Found in the same type of habitat as the previous species but generally distributed; crawling on vegetation in summer, wintering in mud at its base. 10 mm. (See Lilly, 1953.)

RISSOIDAE

Small animals with conical shells, often ribbed and with spiral lines. Operculum horny. Foot very glandular, often with enlarged opercular lobes and metapodial tentacle. Pallial tentacles may also occur. Oesophagus simple; stomach with crystalline style. Larva a veliger.

1. Animal with pallial tentacle on both right and left side (Fig. 7); shell showing reticulated pattern due to interaction of spiral and longitudinal ridges; mouth with inner and outer lips continuous forming a peristome; outer lip thickened with varix . . . . *Alvania* (**a1**)

Animal with pallial tentacle on right side only; shell smooth, or with ribs and spiral lines which rarely intersect; outer lip with or without varix   **2**

Animal without pallial tentacle; shell with ribs and striae on body whorl only, others plain . . . . *Rissoa lilacina* (p. 69)

2. Whorls of shell smooth or ribbed, only occasionally with spiral lines, usually with a labial varix; foot apparently divided into anterior and posterior halves . . . . *Rissoa* (**b1**)

Whorls with spiral lines, occasionally with short adapical ribs, outer lip without varix; foot not apparently divided; metapodial tentacle short
Cingula (**c1**)

a1. Shell with well marked ribs crossed by spiral striae; mouth with thick lips
Alvania crassa (p. 68)

Shell with spiral and longitudinal lines equally pronounced, giving reticulated surface; lips rather thin, outer one with varix   .   *A. punctura* (p. 68)

b1. Shell always with dark comma-shaped mark on apical part of body whorl alongside outer lip (Fig. 54); shell usually smooth but sometimes with ribs . . . . *Rissoa parva* (p. 68)

Shell without such a mark **b2**

b2. Shell without prominent ribs, though stria-like ones occur; spiral striae also present, especially around periphery; body whorl equals two-thirds shell height; whorls 6 . . . . *R. inconspicua* (p. 69)

Shell with prominent ribs, at least on body whorl; outer lip meets inner to form peristome; whorls 7–8 . . . . **b3**

b3. Body whorl equals only about half shell height; ribs well developed except on first few whorls, of which there are 8 . . *R. guerini* (p. 69)

Body whorl equals two-thirds shell height; ribs slight and usually only on body whorl; whorls 7; mouth markedly out-turned *R. membranacea* (p. 69)

e1. Shell with a few spiral striae below periphery on body whorl; other whorls smooth; body whorl somewhat keeled at periphery; mouth pointed where outer lip joins body whorl **c2**

Shell with spiral striae on all whorls; body whorl smoothly rounded . **c3**

c2. Shell with two brown bands, one above and one below periphery of body whorl, one round each older whorl; some shells have an extra band at sutures and at base of body whorl . . . . *Cingula cingillus* (p. 70)

Shell colourless, no bands . . . . . . *C. cingillus* var. *rupestris* (p. 70)

c3. Outer lip meets inner lip to form complete peristome; no umbilical chink; upper parts of whorls usually with short ribs which die out before reaching suture or periphery; mouth equals one-third of shell height

C. semicostata (p. 70)

Outer lip does not meet inner to form peristome; umbilical groove present; ribs absent; mouth equals more than one-third of shell height . . **c4**

c4. Shell uniform buff colour, unbanded, unblotched; umbilical chink by inner lip; spire blunt; metapodial tentacle long, single . *C. alderi* (p. 70)

Shell with brown blotches arranged in bands above and below periphery; umbilical groove by inner lip; metapodial tentacle short, trifid

C. semistriata (p. 70)

The rissoids are all small gregarious animals abounding in rock pools and rock crevices and under stones in summer, though the overwintering population is much reduced, partly perhaps by downward migration, but mainly by death. The animals are usually active and move freely, often climbing up and down mucous strings which originate in the posterior pedal gland which opens about the middle of the sole of the foot. They feed mainly on algae and vegetable detritus.

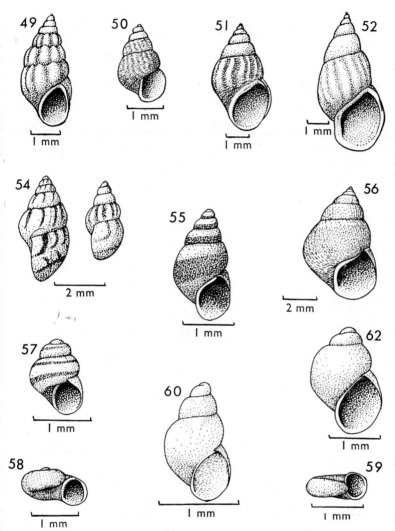

Fig. 49, *Rissoa guerini*; Fig. 50, *Rissoa inconspicua*; Fig. 51, *Rissoa lilacina*; Fig. 52, *Rissoa membranacea*; Fig. 54, *Rissoa parva*, smooth and ribbed variety, drawn to show comma-shaped pigment mark by outer lip; Fig. 55, *Barleeia rubra*; Fig. 56, *Assiminea grayana*; Fig. 57, *Cingulopsis fulgida*; Fig. 58, *Skeneopsis planorbis*; Fig. 59, *Omalogyra atomus*; Fig. 60, *Rissoella diaphana*; Fig. 62, *Rissoella opalina*.

Alvania crassa (Kanmacher) (Fig. 47)

This species may be recognized by the thick peristome, the well developed ribs crossed by spiral striae without producing any reticulate pattern. Shell white. Body clear white. Foot with opaque white V-shaped mark anteriorly. Under stones or on weeds at LWST and below. Scarce everywhere, but less so in south than north. 3 mm.

Alvania punctura (Montagu) (Figs 7, 48)

Recognizable by the rather delicate decussation on the shell surface in which spiral and vertical elements are about equally developed. Body dark. Foot without white V-shaped mark. Among bryozoans and *Laminaria*. Shell yellow-white. Not uncommon locally. 3 mm.

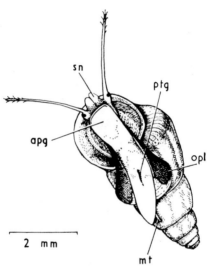

Fig. 53. *Rissoa parva*, living animal seen from below. *apg*, anterior pedal gland; *mt*, metapodial tentacle; *opl*, operculigerous lobe; *ptg*, opening of posterior pedal gland to sole of foot; *sn*, snout. From *British Prosobranch Molluscs* by Fretter and Graham.

Rissoa parva (da Costa) (Figs 53, 54)

The comma-shaped mark by the apical part of the outer lip ensures rapid identification of this species. Two forms exist, one with a smooth, ribless shell (var. *interrupta*), the other with ribs. The variety appears now to be commoner. Shell light brown with chocolate streaks. Animal pale with dark blotches especially near operculum. Abundant in summer amongst corallines and fine weeds in pools, also under stones and in rock crevices. 3–4 mm.

Rissoa inconspicua Alder (Fig. 50)

Shell without peristome, decorated with fine ribs and spiral lines which can usually be seen on all whorls. Outer lip is thin and not out-turned particularly in the area where it adjoins the body whorl. Shell yellowish often with pink apex. Animal white with dark longitudinal lines on sides. On hard bottoms, usually sublittoral; locally not uncommon. 2 mm.

Rissoa guerini Récluz (Fig. 49)

Shell has prominent straight ribs, except on apical turns of spire, which are not crossed by spiral striae though fine lines can be seen in the hollows between ribs. Mouth with prominent peristome, thickened outer lip slightly out-turned. Shell yellowish sometimes with reddish streaks, the ribs being pale and there is often a red colour within outer lip. Animal pale. On algae (*Codium*) or *Zostera* or under stones LW or below; south and west coasts only; locally not uncommon. 4 mm.

Rissoa membranacea (Adams) (Fig. 52)

The diagnostic characters of this species are the distinct out-turning of the lips, the shallow ribbing, largely confined to the body whorl and the height and breadth of that part of the shell. Shell whitish, often darker round the mouth. Animal cream coloured with dark purple sides to the foot. Traditionally associated with *Zostera*, but also occurring on other seaweeds, LWST and below; local, though not common in south. 7 mm.

Rissoa lilacina Récluz (Fig. 51)

Body whorl bears prominent sinuous ribs with distinct spiral striae; other whorls plain. Mouth with distinctly thickened peristome turning out a little all round. Shell whitish, sometimes with longitudinal brown streaks, apex orange and inside of outer lip violet. Animal yellow with dark stripes; orange line along tentacles. Amongst weeds LWST and below, commoner in north. 4 mm.

Cingula cingillus (Montagu) (Fig. 44)

Recognizable by rather regularly conical shell marked by spiral brown bands; spiral striae on basal half of body whorl, which is also slightly angulated peripherally. Mouth pointed where outer lip meets body whorl. Common in silty crevices or under stones from about the mid-level of the beach downwards. 4 mm.

A variety, *C. c. rupestris*, differing from normal by being colourless and in having an unbanded shell is common in rock crevices.

Cingula semicostata (Montagu) (Fig. 45)

This shell gives the appearance of being rather long and slender, marked by incised spiral lines and often having short ribs confined to the apical parts of the whorls; mouth rather small and no umbilicus. Shell white, often very grubby. Animal white. Common in summer under stones, amongst algae and corallines and in silty crevices, from about LW downwards. 3 mm.

Cingula alderi (Jeffreys) (Fig. 43)

A small, unbanded shell with short, blunt spire, decorated with shallow spiral striae. Shell buff. Animal pale yellow with darker blotches on dorsal surface and sides of snout. LWST and below in S.W. England and west coast of Scotland; not common. 1–2 mm.

Cingula semistriata (Montagu) (Fig. 46)

Shell marked by a series of brown blotches which lie one series above, one below the periphery. Spiral lines distinct. Umbilical groove well developed. Background colour of shell yellowish-white; body similar. Abundant in summer under stones and on rock under weed especially where there is silt, at LW and below, on south coasts; much less common in north. 2–3 mm.

These represent the eleven species of rissoid most likely to be encountered by the collector with access to shore and some offshore dredgings. There are more than a dozen other kinds of rissoid recorded from British coasts: the probability of encountering these is very slight and recourse must then be taken to such treatises as Jeffreys' *British Conchology*, Vol. **4** (1867).

BARLEEIDAE

Shell conical, unsculptured. Mouth without labial rib. Operculum horny, concentric. No free larva.

This family contains but the one species, *Barleeia rubra* Adams (Fig. 55). The trivial name derives from the dark red colour of the shell and operculum. The shell is conical, sometimes, in paler varieties, showing one or two broad bands encircling the whorls. Body white with dark lines except for the opercular lobes of the foot which are purple. The animals occur on weeds and in rock pools at low water in S.W. England and Ireland. 3 mm.

ASSIMINEIDAE

Shell small, conical, usually smooth. Animal has no ctenidium though the osphradium persists, short foot without median groove. Oesophagus with gland. Nervous system concentrated.

There is only one British assimineid, *Assiminea grayana* Fleming (Figs 36B, 56), a minute animal recorded only high up on beaches between Kent and the Humber, at the base of grass and sedge plants. It may be recognized by the short tentacles carrying eyes at their tips, the bilobed snout and the ciliated groove running from the mantle cavity down each side of the foot. There are two ciliated ridges in the mantle cavity, one on the mantle skirt, one on the dorsal body wall, which maintain a respiratory current. 5 mm.

CINGULOPSIDAE

Shell like that of a rissoid; operculum horny. Animal without jaws; oseophageal glands present; no crystalline style. No penis; female duct with separate vaginal and oviducal openings. Larval stage suppressed.

A family recently created (Fretter and Patil, 1958) to receive the single species *Cingulopsis fulgida* (Adams) (Fig. 57) to emphasize the very great differences (enumerated above) which separate it from the rissoids with which it was previously classified. The animals occur in rock pools at mid-tide level and below on the more delicate weeds, on which they feed; they also eat detritus. Like many rissoids they clamber about on mucous threads spun from the prominent glands on the pedal sole. Common in summer, much less so in winter; occurs on south and west coasts. 1 mm.

SKENEOPSIDAE

Shell very small, transparent, smooth, with very low spire so that there is a wide umbilicus; mouth round. Operculum present, polygyrous. Shell often covered with algal filaments. Tentacles long. Foot with opening of mucous gland in middle of sole, but no posterior groove. Ctenidium present in mantle cavity Reproductive system specialized. No free-swimming larval stages.

Only one species, *Skeneopsis planorbis* (Fabricius) (Fig. 58) with the character of the family, is found in the British fauna. It occurs amongst small weeds on the lower part of rocky coasts and below. Common in summer, rare in winter. 1 mm (See Fretter, 1948.)

OMALOGYRIDAE

Shell very small, transparent, smooth or with transverse sculpturing; whorl all in one plane so that there is no spire. Mouth circular; snout bifid forming two semicircular head lobes; no tentacles. Foot with median gland aperture Ctenidium and osphradium lost. No free-swimming larval stages.

Only one species, *Omalogyra atomus* (Philippi) (Figs 59, 61B), is likely to be found. It occurs in pools in the lower half of the beach on all rock shores particularly where *Ulva* grows. It often spins a rope of mucus and climbs up and down the pool on this. It is often abundant in summer though scarce in winter About 1 mm *across* shell. (See Fretter, 1948.)

A second species *Ammonicera rota* (Forbes and Hanley) is recorded in the British fauna from similar situations at LWST in S.W. England and S. Ireland It is distinguished by a keel at the periphery of the body whorl and some mark ings running across the whorls. It seems to be very rare though it may just have been overlooked because of its size. 0·5 mm.

RISSOELLIDAE

Shell very small, transparent, smooth, with short spire loosely wound so that it has an umbilicus. Operculum present. Foot with prominent opening of the posterior mucous gland centrally placed connected to hind tip of foot by groove. Ctenidium lost. No free larval stages.

Shell with body whorl about two-thirds of total height; tentacles of animal short snout; bifid forming two triangular lobes . . . *Rissoella diaphana*

Shell with body whorl about five-sixths of total height; tentacles bifid (Fig. 61A); snout simple . . . . *R. opalina*

These are minute prosobranchs which abound in rock pools in the lower half of the beach in summer-time; overwintering population low. They are feeders on vegetable detritus and, presumably because of their specialized habitat, have suppressed free-swimming larval stages. (See Fretter, 1948.)

Rissoella diaphana (Alder) (Fig. 60)

Recognizable by characters shown in key. The shell allows some internal anatomy to be seen. The egg capsules are attached to red or green weeds in pools. Common, especially in summer. 1–2 mm.

Rissoella opalina (Jeffreys) (Figs 61A, 62)

As in *R. diaphana* some internal anatomical characters may be seen by virtue of the transparent shell. The egg capsules are attached to red or green weeds in pools. Common, especially in summer but not found on east coast. About 2 mm.

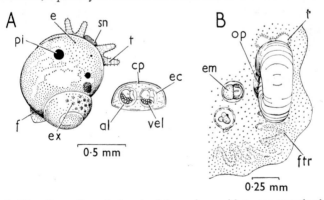

Fig. 61. **A**, *Rissoella opalina*, whole animal from above with an egg capsule alongside, **B**, *Omalogyra atomus*, whole animal from above, on weed with two egg capsules. *al*; albumen; *cp*, egg capsule; *e*, eye; *ec*, egg covering; *em*, embryo; *ex*, excretory material in digestive gland; *f*, foot; *ftr*, feeding track; *op*, operculum; *pi*, pigment in mantle skirt; *dn*, snout; *t*, cephalic tentacle; *vel*, veliger larva. From *British Prosobranch Molluscs* by Fretter and Graham.

TURRITELLIDAE

Shell tower-shaped, with many whorls, usually with spiral ridges; mouth small; operculum horny, often with marginal processes. Mantle edge bears tentacles. Mantle cavity and ctenidium elongated, often adapted for ciliary food collecting. Radula small, oesophagus without glands, stomach with crystalline style. Genital ducts open, male aphallic. Larva a veliger.

There is only one British species, *Turritella communis* Risso (Fig. 63), sometimes known as an auger or screw shell. It shows most of the family characters. It lives at shallow depths on muddy bottoms, partly buried and maintains a ciliary current through the mantle cavity not only for respiratory purposes but also for food collection. Particles strained out of the water current are carried across the pallial floor in a groove to the mouth. The animals are lethargic and hardly move. Sperm are filtered out of the current entering the mantle cavity of the females and passed to the open genital ducts where fertilization occurs. The animals are gregarious and so are locally abundant. 40 mm. (See Graham, 1938; Yonge, 1946.)

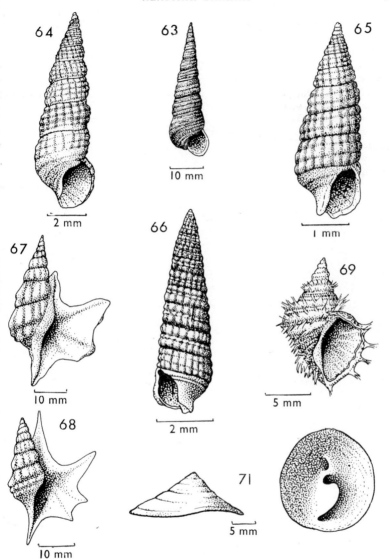

Fig. 63, *Turritella communis*; Fig. 64, *Bittium reticulatum*; Fig. 65, *Cerithiopsis tubercularis*; Fig. 66, *Triphora perversa*; Fig. 67, *Aporrhais pespelicani*; Fig. 68, *Aporrhais serresiana*; Fig. 69, *Trichotropis borealis*; Fig. 71, *Calyptraea chinensis*, in lateral and ventral views.

F

CAECIDAE

Shell small; starts with normal spiral shape but with growth changes to open spiral and the initial turns then break off and the hole is plugged with a calcareous cap. Nervous system concentrated, visceral ganglion lost. Sexes separate, female with separate openings to oviduct, "uterus" (producing albumen for eggs) and nidamental gland, secreting egg capsule. Larva a veliger.

There are two species of this highly modified family in the British fauna *Caecum glabrum* (Montagu) and *C. imperforatum* (Kanmacher). Both are dredged from sandy bottoms, though rarely. The former has a smooth shell and a bulging operculum, the latter has a shell marked with distinct rings and a flat operculum. *C. glabrum* is smaller, about 2 mm; *C. imperforatum* measures 3 mm.

CERITHIIDAE

Shell tower-shaped, with many whorls, usually sculptured. Operculum horny with few turns. Radula short. Oesophageal glands present. Genital ducts open. Larva a veliger.

The only British species in this family is *Bittium reticulatum* (da Costa) (Fig. 64) which is not common, but because of its gregarious habit, several may be found simultaneously under stones, or in crevices at LWST or below on rocky shores. It is limited to south-west coasts. Occasionally burrowing in sandy mud. 10 mm.

CERITHIOPSIDAE

Shell small, tower-shaped, marked with spiral rows of tubercles; mouth small, siphonal canal very short. Radular teeth clawed. Salivary glands lie one behind the other. Oesophageal gland present with glandular diverticulum. Genital ducts open, males aphallic. Larva a veliger.

There are several genera of British cerithiopsids, but only one species is likely to be found, *Cerithiopsis tubercularis* (Montagu) (Fig. 65), common on west and south-west coasts where it occurs in coralline pools, rock crevices and under boulders in lower parts of the beach. Shell chestnut brown, almost parallel-sided at base, tapering to a point only in the region of the first 7 whorls. Browses on sponges. 7 mm. (See Fretter, 1951*a*.)

TRIPHORIDAE

Shell sinistral, tower-shaped, marked with spiral rows of tubercles; mouth small, siphonal canal very short. Marginal radular teeth with long comblike processes. Oesophageal gland connects to oesophagus by duct at posterior end. Genital ducts open, males aphallic. Larva a veliger.

This family contains only one British species, *Triphora perversa* (L.) (Fig. 66), which may be instantly recognized as the only regularly sinistral British proso-branch. It is chestnut brown in colour, occurring in rock crevices in and just above the laminarian zone, and may be dredged from hard bottoms on south and west coasts. Locally common, feeding on monaxonid sponges. The sinistrality is true, affecting all parts of shell and body, so that they are the mirror image of those of other prosobranchs. 8 mm.

APORRHAIDAE

Shell tall, with ribs; body whorl also with tubercles: outer lip with palmate extension joined basally to siphonal canal. Operculum set across foot. Genital ducts open. Larva a veliger.

There are two British species, *Aporrhais pespelicani* (L.) (Fig. 67) and *A. serresiana* (Michaud) (Fig. 68). The latter has been recorded only from deep water off Arran and the Shetland Isles; the former occurs fairly commonly on sandy mud or gravel off all coasts. It spends at least part of its time buried shallowly in the mud; the expansion of the lip forms the roof of an area round the animal's head in these circumstances and protects two siphons which lead water to and from the mantle cavity. The male has a ciliated area on the floor of the right-hand side of the mantle cavity at its opening and the females have a ciliated tract in the same position which is extended on to the foot. 30 mm. (See Yonge, 1937.)

TRICHOTROPIDAE

Shell with short spire, covered with periostracum carrying rows of bristles. Horny operculum. Animal with short grooved proboscis like that of a *Capulus*, turned to right side of head. Oesophageal gland well developed. Nervous system concentrated. Protandrous hermaphrodites. No free-swimming larvae.

Trichotropis borealis Broderip and Sowerby (Figs 2, 69)

Shell with spiral ridges on which lie the rows of periostracal bristles. The only unusual feature of the body is the proboscis. Dredged from hard bottoms in north. 12 mm.

CAPULIDAE

Shell cap-shaped with marked periostracum. No operculum. Shell muscle horseshoe-shaped. Ventral lips drawn out to form proboscis. Larva a veliger.

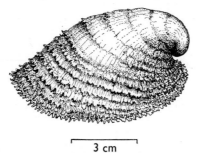

3 cm

Fig. 70. *Capulus ungaricus.* From *British Prosobranch Molluscs* by Fretter and Graham.

Only one British species *Capulus ungaricus* (L.) (Fig. 70), the Hungarian cap shell. The shell is cap-shaped with a backwardly turned, slightly coiled apex, and is covered with a thick brown periostracal layer which projects at the mouth to give a short fringe. The shell is marked with spiral ridges except near the mouth. The animal is distinguished at once by the presence of a long grooved proboscis which cannot be retracted. It lives commonly on the shells of bivalves, especially scallops, and feeds by thrusting the proboscis into the mantle cavity of its host and licking up the mucus and fine particles collected by its host. It has also been found on *Turritella communis* feeding in the same way. *Capulus* may also be found on rocks. Dredged from shallow water off all coasts, on shelly rocky bottoms. 12 mm. (See Yonge, 1938.)

CALYPTRAEIDAE

Shell conical or flattened hemisphere, with internal septum. Shell muscle single with small oval attachment. Mantle cavity enlarged. Neck with lateral lobes. Ciliary feeding animals; oesophageal glands absent. Protandrous hermaphrodites. Larva a veliger.

Shell circular in outline, smooth; animal solitary . . *Calyptraea chinensis*

Shell oval in outline, rough; animals occur one on top of the other, in chains
Crepidula fornicata

A small family of limpets which live at LW and below. They are ciliary feeders and have a crystalline style in the stomach. They are protandrous hermaphrodites. In *Calyptraea* the small male is temporarily associated with the larger female; in *Crepidula* he settles permanently on a female and the animals form chains (Fig. 73) of up to 12 individuals, small males at one end, animals changing sex in the middle and females at the base. The lowest part of the chain is usually an empty shell.

Calyptraea chinensis (L.)—Chinaman's hat (Fig. 71, p. 75)

Shell a more or less regular cone, often showing slight spiral coil at apex; surface smooth, mouth nearly circular, though in side view shell is curved to fit the object to which it adheres; internal shelf with C-shaped free edge. Common on stones and shells from stony grounds in shallow water on west coast, becoming intertidal (LW only) in sheltered situations in S.W. England. 10 mm.

Crepidula fornicata (L.)—Slipper limpet (Figs 72, 73, 74)

Shell oval, showing some spiral coiling, surface marked with growth lines, mouth oval; free edge of internal shelf almost straight. Animals occurring in chains which abound in sheltered areas in a few metres depth and may be rolled ashore. Introduced from North America with imported oysters to Essex river and Mersey estuary, though it died out in the latter situation. It has now spread from Essex south and west to Wales and north to Scotland. The lower (female) members of the chain brood eggs under the foot during the breeding season. In Essex a serious pest of oyster beds; the limpets may form a layer inches deep over the oysters, using up their food, and smothering them with faeces and pseudofaeces. Up to 50 mm.

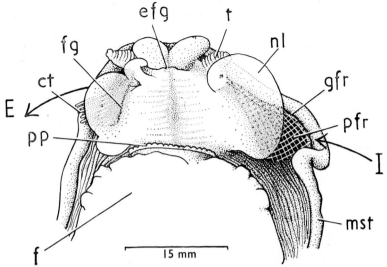

Fig. 72. *Crepidula fornicata.* Ventral view of the anterior half of the body. Arrows show direction of water flow, *I*, into the mantle cavity, *E*, out of it. *ct*, ctenidium; *efg*, anterior end of food groove; *f*, foot; *fg*, food groove seen through neck lobe; *gfr*, groove from pallial filter to food pouch; *mst*, mantle skirt; *nl*, neck lobe; *pfr*, pallial mucous filter; *pp*, propodium; *t*, cephalic tentacle. From *British Prosobranch Molluscs* by Fretter and Graham.

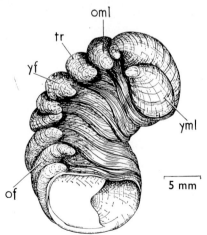

Fig. 73. *Crepidula fornicata.* Chain of animals. *of*, old female; *oml*, old male; *tr*, transitional form; *yf*, young female; *yml*, young male. From *British Prosobranch Molluscs* by Fretter and Graham.

LAMELLARIIDAE

Shell partly or wholly covered by mantle, thin and polished. No operculum. Two columellar muscles. Gut with extensile proboscis. Oesophageal gland well developed with lamellae. Female with ventral pedal papilla, by means of which the egg capsules are deposited in holes bitten out of the tunicate colonies on which the adults feed. Larva an echinospira, with double shell.

1. Shell only temporarily covered by mantle lobes   .   .   .   .   (*Velutina*) **2**

Shell internal, completely and permanently covered by mantle which has anterior notch used as siphon   .   .   .   .   (*Lamellaria*) **3**

2. Shell clearly marked with spiral lines; its mouth circular; animal pale in colour, white-yellow   .   .   .   .   *Velutina velutina*

Shell without spiral lines; its mouth oval, animal bright orange

V. plicatilis

3. Animal about 20 mm long, purplish-grey in colour with lighter flecks

Lamellaria perspicua

Animal about 10 mm long usually sandy in colour, always with some black flecks.   .   .   .   .   *L. latens*

These animals are specialized for feeding on tunicates and their colours match those of the prey. *Lamellaria* is often mistaken for a dorid, but it has no dorsal circlet of gills, and on being examined ventrally the ctenidium may be seen in the mantle cavity and two long tentacles with prominent basal eyes are visible.

Velutina velutina (Müller) (Fig. 75)

Shell covered with brown periostracum; 3 whorls. Mantle edge thickened so as to limit entry to mantle cavity to ventral notch anteriorly on left and egress to ventral notch posteriorly on right. Lives on the ascidian *Styela coriacea*. Fairly common LW and below, commoner in north. Up to 20 mm.

V. plicatilis (Müller)

Feeds on hydroids. Not common, northern. About 12 mm.

Lamellaria perspicua (L.)
L. latens (Müller)

These two animals are of similar habit and habitat, found under stones at LW and below. They feed especially on the tunicates *Leptoclinum* and *Polyclinum*. *L. perspicua* (10 mm) grows to about twice the size of *L. latens* (5 mm) but animals of the latter species always seem to have some flecks of dark pigment in their skin and to be sandy in colour, whereas *L. perspicua* lacks this and is usually a grey-lilac shade. (See McMillan, 1939.)

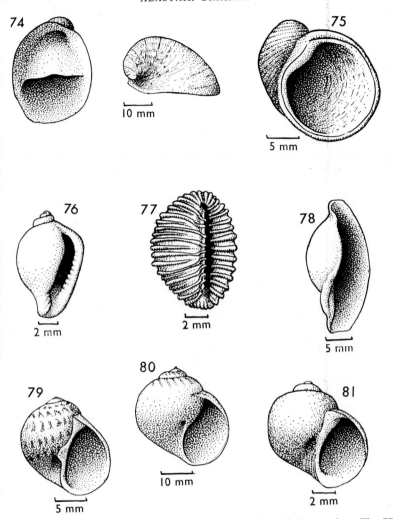

Fig. 74, *Crepidula fornicata*; Fig. 75, *Velutina velutina*; Fig. 76, *Erato voluta*; Fig. 77, *Trivia monacha*; Fig. 78, *Simnia patula*; Fig. 79, *Natica alderi*; Fig. 80, *Natica catena*; Fig. 81, *Natica montagui*.

ERATOIDAE

Shell with very short spire or convolute, polished; mouth long and narrow. No operculum. Shell muscle double. Mantle edge can extend over shell. Gill curved osphradium curved. Pedal ganglia ovoid. Larva an echinospira, with doubl shell.

1. Shell with short spire   .   .   .   .

Shell convolute, no spire visible.   .   .   .   .

2. Shell slightly ribbed, base regularly oval in outline; outer lip curved upward towards tip of spire   .   .   .   .   *Trivia* juvenil

Shell smooth, base narrowed to give a harp-shape in outline; outer lip curve downwards towards base of shell   .   .   .   .   *Erato volut*

3. Shell with 3 dark spots on side opposite mouth   .   .   .   *Trivia monach*

Shell without spots.   .   .   .   .   *T. arctic*

The eratoids were previously classified as cypraeids but are now separated from them on the basis of a number of anatomical features and on their possession o an echinospira larva. In *Trivia* spp. the shell of the young animal shows the olde spiral turns; at maturity the last whorl grows over and conceals them, producing the type of shell known as convolute.

Erato voluta (Montagu) (Fig. 76)

Shell harp-shaped, covered over by lateral mantle folds when the animal i active. The animal eats compound ascidians (*Botryllus*, *Botrylloides*). Dredged i shallow water; commoner in south. 10 mm.

Trivia monacha (da Costa) (Fig. 77)

The common British cowrie found under stones and in crevices of rocks a LWST and below. The shell is normally covered by lateral mantle folds which are brightly and variously coloured. Cowries eat compound ascidians (*Diplosoma Botryllus*, *Botrylloides*) and lay egg capsules in holes bitten out of their common test. 10 mm.

T. arctica (Montagu)

As the last except for the absence of spots on shell. Rarely littoral, normally dredged.

CYPRAEIDAE

Shell convolute and polished; mouth long and narrow, drawn out at each end. No operculum. Shell muscle double. Mantle edge can extend over shell. Gill curved, osphradium triradiate. Pedal ganglia elongated into pedal cords with scalariform connexions. Larva a veliger.

A very extensive family of gastropods including the animals popularly called cowries. Only one genus *Simnia* occurs in Britain, most of the cowries being limited to tropical areas. *Simnia patula* (Pennant) (Fig. 78) may be dredged from south-western waters only on colonies of Dead men's fingers (*Alcyonium digitatum*) on the polyps and flesh of which the mollusc feeds, biting pieces with radula and a pair of jaws. 25 mm.

NATICIDAE

Shell globular, with low spire and expanded body whorl, umbilicate, smooth. Operculum horny. Foot with expanded propodium which covers the head when the animal creeps, and with lateral folds which cover sides and posterior part of shell. Accessory boring organ on ventral lip. Oesophageal gland well developed. Nervous system concentrated. Larva a veliger.

1. Shell without reddish-brown zigzag streaks; outer lip thick, especially where it meets body whorl; a ridge from the inner lip runs into the umbilicus; not littoral . . . . *Natica montagui*

Shell with reddish-brown zigzag streaks; outer lip thin, no ridge entering umbilicus; littoral or sublittoral **2**

2. Shell with one row of reddish-brown streaks on upper part of each whorl; umbilicus not blocked by inner lip; outer lip curves sharply in to meet body whorl more or less at right angles; littoral . . . . . . *N. catena*

Shell with several rows of reddish-brown or dark brown streaks on body whorl; umbilicus partly closed by flat extension of inner lip; outer lip slopes up to meet periphery of body whorl; LWST downwards *N. alderi*

Natica alderi Forbes (Fig. 79)

Recognizable by the pigment pattern on the shell, the half closure of the umbilicus by the inner lip and by the sloping outer lip as it approaches the body whorl. In sandy bays, and dredged from sandy bottoms LWST downwards. 15 mm.

Natica catena (da Costa) (Fig. 80)

Characterized by pigment pattern, open umbilicus and incurving outer lip. In sandy bays at LWST and below. 30 mm.

Natica montagui Forbes (Fig. 81)

Not apparently littoral, but dredged on sandy, gravelly bottoms. Distinguished by lack of pigmented streaks, thickness of outer lip, the umbilical ridge, white band frequently round the suture, which is deeper than in other species. 10 mm.

Two other species *N. fusca* Blainville and *N. pallida* Broderip and Sowerby are dredged too rarely to need more than mention.

SCALIDAE

Shell tower-shaped, with many whorls, usually with ribs; mouth round; operculum horny. Long proboscis present; radula ptenoglossan; salivary glands tubular, their opening (outside mouth) armed with a stylet; oesophagus without gland; stomach simple. Animals hermaphrodite and aphallic. Spermatozeugmata are produced. Larva a veliger.

There are several British species of this family but the only one likely to be encountered is the wentletrap, *Clathrus clathrus* (L.) (Fig. 82), which may be occasionally met on muddy or sandy rocky shores particularly in spring; it has then probably come from a sublittoral position to spawn, laying a long string of small triangular capsules. The mode of life is unknown, but if it is like that of some of its relatives it probably preys on sea anemones or other coelenterates. Shell colourless or pale fawn with very rounded whorls, just meeting, and buttressed with flat-sided projecting ribs. Animal often exudes purple dye. 20 mm.

IANTHINIDAE

Shell thin, smooth, spire low; operculum absent. Foot with lateral folds and secreting a float formed of bubbles of air trapped in mucus. Tentacles bifid, eyes invisible. Radula ptenoglossan. Animal secretes purple dye. Protandrous hermaphrodites, male aphallic producing spermatozeugmata. Larva a veliger.

There is one British genus, *Ianthina*.

1. Shell with tall spire, mouth angulated near inner lip **2**

Shell with low spire, mouth rounded near inner lip   .   .   .   .   *I. pallida*

2. Shell surface smooth .   .   .   .   *I. janthina*

Shell surface with prominent V-shaped furrows   .   .   .   .   .   *I. exigua*

The Ianthinidae are adapted for gregarious planktonic life. The shell is thin, purple on the apical part and pale basally which helps (as it floats upside down) to make it invisible from above and below. The mollusc keeps itself at the surface of the sea by means of a float of trapped air bubbles, to which *I. pallida* and *I. exigua* attach their egg capsules. The other species is viviparous, the eggs being fertilized in the ovary by spermatozeugmata which swim up the female ducts. All are carnivores, feeding mainly on the planktonic siphonophores *Porpita* and *Velella*. The animals are occasionally drifted and stranded on the coasts of S.W. England, S. Wales and S. Ireland, especially in summers with long periods of westerly winds. (See Laursen, 1953; Wilson and Wilson, 1956.)

Ianthina janthina (L.)

Shell smooth. Viviparous. 20–30 mm.

Ianthina exigua Lamarck (Fig. 81a)

Shell with V-shaped furrows. Oviparous. 10–15 mm.

Ianthina pallida Thompson

Shell globose with rather rough surface and central keel. Oviparous. 20 mm.

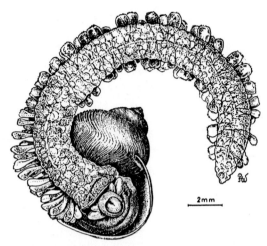

Fig. 81a. *Ianthina exigua* Lamarck in its floating position as viewed from above (after Laursen, 1953).

EULIMIDAE

Shell with long spire of many whorls, smooth and highly polished. Radula absent. Gut with long proboscis with pump at base; oesophageal glands absent. Nervous system concentrated. Reproductive system with open ducts. Sexes separate, though males may be uncommon. Larva a veliger.

1. Shell white 2

Shell with two brown bands round middle of body whorl and maybe also a fainter third below suture . . . . *Eulima trifasciata*

2. Shell forming a straight cone; animal associated with *Spatangus*

Balcis alba

Shell forming a skew cone; animal associated with *Antedon* . *B. devians*

The family contains animals which have become ectoparasitic on echinoderms to which they attach themselves with the aid of secretions from hypertrophied pedal glands. They penetrate the tissues of the host with a very long proboscis, the base of which has a fold or pseudopallium on it, and suck food by means of a pump made by modification of the buccal cavity. Radula and jaws have been lost and the whole gut structure secondarily simplified. (See Fretter, 1955.)

Eulima trifasciata (Adams) (Fig. 83)

Shell characterized by coloured bands. Body white. Tentacles rather long and close together with eyes in line with their base. Under stones on rocky beaches and dredged on gravelly bottoms. Mode of life unknown. Not common. 8 mm.

Balcis alba (da Costa) (Fig. 84)

Shell white and smooth, sometimes with a slight curvature. Body white with yellow speckles sometimes forming a V-shape on snout, and a ring round each eye. Not littoral. Dredged from areas in which the heart-urchin, *Spatangus purpureus*, lives; may feed on this. Off all coasts. 18 mm.

B. devians (Monterosato (Fig. 85)

Recognizable by reason of the curved spire. Otherwise like shell of *B. alba*. Body flecked with colour, usually red; spots on body and continuous pigment on base of tentacles. Usually dredged but may be found in rock pools at LWEST. Ectoparasitic on *Antedon*. Locally common. Males rare or not found. 4 mm.

There are 6 other species of *Balcis* and 2 other species of *Eulima*. These are all extremely rare and not at all likely to be found, but see Forbes and Hanley (1849–53) or Jeffreys (1862–69).

STILIFERIDAE

Shell rather globular but with narrow style-like apical region. No operculum. Animal with proboscis carrying a fold known as the pseudopallium. Hermaphrodite.

Only one British species, *Pelseneeria stylifera* (Turton) (Fig. 86), has been recorded (from *Psammechinus miliaris* off Plymouth, and *Echinus esculentus* elsewhere). It adheres to the surface of the echinoid, creeping among the spines and browsing on the tissues on which it also deposits its spawn. 3–4 mm.

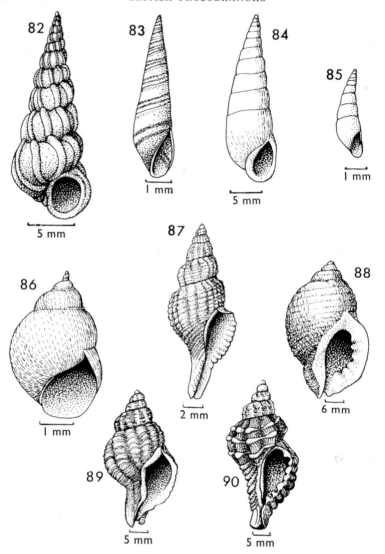

Fig. 82, *Clathrus clathrus*; Fig. 83, *Eulima trifasciata*; Fig. 84, *Balcis alba*; Fig. 85, *Balcis devians*; Fig. 86, *Pelseneeria stylifera*; Fig. 87, *Trophon muricatus*; Fig. 88, *Nucella lapillus*; Fig. 89, *Urosalpinx cinerea*; Fig. 90, *Ocenebra erinacea*.

MURICIDAE

Shell with siphonal canal and often with ribs or other outgrowths. Operculum with terminal nucleus. Animal with proboscis, broad foot with accessory boring organ; females have a ventral pedal gland. Radula rachiglossan. Accessory salivary glands present; oesophageal valve and separate oesophageal gland present; stomach simple; anal gland occurs. Nervous system very concentrated, though visceral ganglia are not included in the nerve ring. Hypobranchial gland secretes a purple secretion. Eggs laid in horny capsules fastened to substratum. Veliger stage often suppressed.

1. Shell without ribs; siphonal canal separated from rest of shell by deep spiral
   groove .   .   .   .   *Nucella lapillus*

   Shell ribbed; no deep spiral groove **2**

2. Shell turreted, siphonal canal long, equal to height of mouth; not littoral
                                                            Trophon muricatus

   Shell not turreted, siphonal canal short, equal to half height of mouth at
   most **3**

3. Siphonal canal open; many small spiral ridges over ribs; inner side of outer
   lip smooth; E. Anglia only   .   .   .   .   *Urosalpinx cinerea*
   Siphonal canal closed (in adult animals); a few rather broad spiral ridges
   over ribs; inner side of outer lip toothed   .   .   .   .   *Ocenebra erinacea*

The muricids are predators capable of boring the shells of other molluscs to reach the soft parts. This is carried out mechanically by the radula aided by some chemical process still not precisely known effected by the accessory boring organ, kept, when not in use, in a pit in the anterior part of the sole of the foot (Fig. 8). They will also feed on prey, if it is defenceless enough, simply by pushing the proboscis between the valves of a lamellibranch or barnacle shell.

Nucella lapillus (L.)—Dog whelk (Fig. 88)

Shell marked with spiral lines but no ribs; body whorl occupies most of spire. In young animals the outer lip is sharp and thin; at maturity it becomes thickened and grows internal teeth. The colour is very variable, commonly ashy, but frequently with chocolate spiral bands; black, white and yellow shells also occur. The presence of brown pigment has been associated with feeding on mussels but this may not necessarily be so. Animal cream-coloured. The animals occur on all rocky coasts except very exposed ones, abundantly wherever barnacles and mussels are found, usually collected under stones and in rocky crevices from HWNT downwards. Egg capsules are also found in similar situations. 30 mm.

Trophon muricatus (Montagu) (Fig. 87)

Shell tall and conical, with upper part of each whorl flattened to give a turreted appearance. Sculpture: ribs crossed by spiral grooves in peripheral region of body whorl and in all whorls of the spire. Siphonal canal elongated, siphon usually extending beyond its tip. Animal white. Dredged on muddy sand on west coasts; never common but more frequent in the south. 15 mm.

Urosalpinx cinerea (Say)—American oyster drill (Fig. 89)

Shell moderately tall and conical; whorls convex. Ribs crossed by fine spiral ridges; siphonal canal short and bent towards left. Shell pale buff. This animal is limited to oyster beds at LWST and below on the coasts of Essex. It was introduced from America at the end of the nineteenth century. It is common, as are its egg capsules, attached to stones and shells. 25 mm.

Ocenebra erinacea (L.)—Rough tingle (Figs 8, 90)

Shell rather irregularly sculptured with a small number of outstanding ribs crossed by moderately stout spiral ridges. In young animals the siphonal canal is open along its length but at maturity the canal closes except at the tip; at the same time the outer lip becomes thickened and internal teeth appear. Shell yellowish or white; animal yellow or cream with white flecks. On rocks and under stones on sheltered rocky shores at LW and below. Egg capsules on stones and shells. Moderately common in south and west but much less so in north. 30 mm.

BUCCINIDAE

Shell tall, conical, often sculptured with ribs and spiral ridges. Operculum horny with nucleus sometimes terminal, sometimes central. Animal with proboscis, broad foot with ventral pedal gland in females. No accessory boring organ. Radula rachiglossan. No accessory salivary gland present; oesophageal valve and gland reduced; stomach simple; no anal gland. Nervous system concentrated. Eggs laid in horny capsules, usually not attached to substratum. Veliger stage often suppressed.

1. Shell with ribs **2**
   Shell without ribs **3**

2. Large animals (shell height up to 70 mm) with spiral ridges crossing ribs; mouth of shell broad oval, no labial rib . *Buccinum undatum* (p. 96)

   Small animals (shell height about 4–5 mm) with spiral grooves crossing ribs; mouth of shell narrow, outer lip with broad rib
   Chauvetia brunnea (p. 96)

3. Shell with marked periostracum; whorls do not dip much to sutures; no spiral ridge at siphonal canal **4**

   Shell without periostracum; whorls dip distinctly to sutures; pronounced spiral ridge alongside siphonal canal . . . *Neptunea antiqua* (p. 96)

4. Spire more or less flat-sided; periostracum absent from a triangular area where outer lip meets body whorl; sculpture fine spiral lines
   Colus gracilis (p. 96)

   Spire shows whorls dipping to sutures; no bare triangular patch, though there is a linear area devoid of periostracum where outer lip meets body whorl
   C. jeffreysianus (p. 96)

The buccinids are successful animals, more properly carrion feeders than true carnivores, gaining access to dead food in crevices by means of their very extensible proboscis. They do not bore shells.

Buccinum undatum L.—Whelk or buckie (Fig. 94)

The shell is recognizable by the sinuous ribs ("*undatum*") crossed by spiral ridges. The ribs do not extend to the base of the body whorl. Shell greyish white; animal white or cream with numerous irregular black flecks. Operculum horn coloured depressed centrally towards the nucleus. On rocky and sandy shores at LW and below to deep water. Egg capsules fastened to one another to make "sea-wash balls" often cast ashore when empty. 80 mm.

Chauvetia brunnea (Donovan) (Fig. 95)

Shell small with whorls dipping little to sutures. Body whorl with ribs not extending much below periphery. Basal part of body whorl with spiral ridges only. Ribs are crossed by narrow spiral grooves. The shell mouth is narrowly oval and the outer lip bears a prominent labial rib or varix. Shell tan or brown in colour; animal cream with opaque white spots. Occasionally found under stones on rocky shores at and below LWST in S.W. England. 5 mm.

Neptunea antiqua (L.)—Red whelk or buckie (Fig. 93)

Shell resembling that a whelk (*Buccinum*) but devoid of ribs and with only spiral lines for sculpture; the siphonal canal has a large spirally curved ridge on its side. Shell yellow or even reddish in colour; animal cream, sometimes with black speckles. At LWST or more commonly below, on all coasts except south; more frequent in north. Egg masses like those of *Buccinum*. 80 mm.

Colus gracilis (da Costa) (Fig. 91)

Spire of shell rather flat-sided as whorls dip only slightly to sutures. Sculpture in the form of slight spiral lines. Periostracum well developed though absent from triangular area on body whorl where the outer lip joins it. Shell a deep fawn; animal cream. Dredged from muddy sand especially in north; it is rare in south. 70 mm.

Colus jeffreysianus (Fischer) (Fig. 92)

Whorls dip to sutures. Sculpture in the form of spiral lines stronger than those in *C. gracilis*. Periostracum slight, absent from an elongated area near inner lip. Shell yellowish; animal cream. Dredged rarely from sandy grounds or bryozoan beds off S. and W. England only. 50 mm.

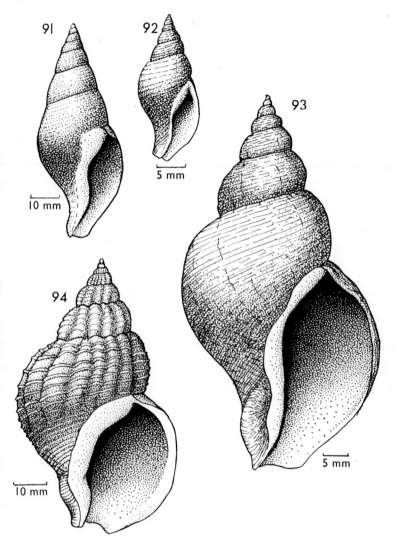

Fig. 91, *Colus gracilis*; Fig. 92, *Colus jeffreysianus*; Fig. 93, *Neptunea antiqua*; Fig. 94, *Buccinum undatum.*

NASSIDAE

Shell of medium size, variously sculptured; siphonal canal short. Foot with two tentacles at hind end; opening of ventral pedal gland in centre of sole in females. Siphon long. Larva a veliger.

1. Shell with convex ribs, whorls dip to sutures   2

   Shell with flat ribs, and reticulated pattern; spire flat-sided
   Nassarius reticulatus

2. Shell buff, with dark chocolate blotch on base; outer lip and its supporting rib dark; littoral and below . . . .   *N. incrassatus*

   Shell pale buff, without basal blotch; outer lip and its rib pale; not littoral
   N. pygmaeus

The nassids are active animals which abound in the sandier parts of rocky beaches where they live on carrion sought out by a lively olfactory sense operating through the mobile siphon. Egg capsules attached to rock.

Nassarius reticulatus (L.) (Figs 3, 98)

Shell conical, slightly cyrtoconoid, marked by a regular reticulation of square areas produced by interaction of flat ribs and spiral ridges. A deep spiral groove separates the siphonal canal from the base of the shell. Outer lip thick with plaits within. Shell chestnut brown. Found at LWST on all rocky beaches, under stones, in crevices especially where silty, and below. 25 mm.

Nassarius incrassatus (Ström) (Fig. 97)

Shell conical, with convex whorls dipping to sutures, marked with elevated ribs crossed by spiral ridges. Deep spiral groove separates siphonal canal from shell. Outer lip as in *reticulatus*. Shell light brown with characteristic black blotch on outside of siphonal canal. In same situations as *reticulatus*. 10 mm.

Nassarius pygmaeus (Lamarck) (Fig. 96)

Similar in general appearance to *N. incrassatus* but without the basal blotch of dark colour and with a pale, almost white, labial thickening. Not littoral, dredged occasionally from sandy bottoms of S.W. England. 10 mm.

CONIDAE

Shell conical, with variable sculpture, with siphonal canal at base for inhalant siphon and frequently an exhalant notch overlying the anus at the point where the outer lip joins body whorl. Operculum present or absent. Radula and anterior part of gut highly modified in relation to predatory habit; poison gland present.

1. Animal with operculum . . . . *Lora turricula* (p. 101)

Animal without operculum **2**

2. Shell marked with ribs only **3**

Shell with ribs and spiral ridges which interact to give reticulated pattern **7**

Shell marked with spiral ridges only . . . . *Philbertia teres* (p. 101)

3. Suture plain **5**

Suture marked by plain or beaded ridge on its basal side **4**

4. Suture with beaded ridge, whorls dipping to sutures; mouth very narrow, siphonal canal tapers . . . . *Mangelia nebula* (p. 102)

Suture with plain ridge, whorls flat-sided; siphonal canal rather wide
M. powisiana (p. 102)

5. Shell not glossy, mouth with obvious anal notch where outer lip joins body whorl, ribs rather broad **6**

Shell glossy, mouth with rather small, shallow anal notch where outer lip joins body whorl, ribs narrow and sharp, 9 on the body whorl; siphonal canal tapers . . . . *M. attenuata* (p. 102)

6. 9–10 ribs on body whorl; mouth narrow; siphonal canal trends to left
M. costulata (p. 102)

7 ribs on body whorl; mouth open; siphonal canal straight
M. coarctata (p. 102)

7. Shell elongated; 14 ribs on body whorl; ribs do not extend to sutures; mouth long . . . . *Philbertia gracilis* (p. 101)

Shell with not more than 12 ribs on body whorl; ribs extend to sutures . **8**

8. 12 ribs and 12 spiral ridges on body whorl; whorls tumid; mouth pear-shaped with distinct anal notch; columella flexuous . . . *P. linearis* (p. 101)

7–9 ribs and about 15 spiral ridges on body whorl; whorls tumid, but flatten as they approach sutures; mouth narrow with not well marked anal notch; columella straight . . . . *Mangelia brachystoma* (p. 102)

H

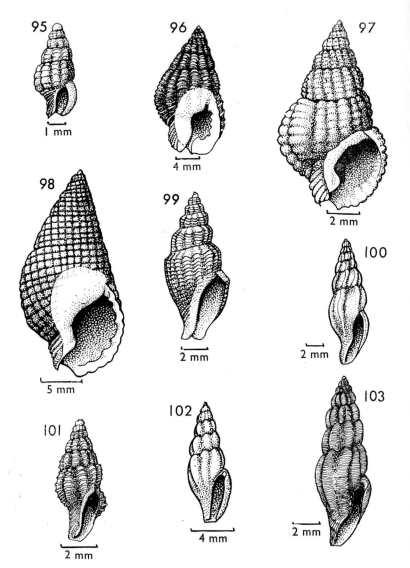

Fig. 95, *Chauvetia brunnea*; Fig. 96, *Nassarius pygmaeus*; Fig. 97, *Nassarius incrassatus*; Fig. 98, *Nassarius reticulatus*; Fig. 99, *Lora turricula*; Fig. 100, *Mangelia attenuata*; Fig. 101, *Mangelia brachystoma*; Fig. 102, *Mangelia coarctata*; Fig. 103, *Mangelia costulata*.

The Conidae are predominantly tropical in distribution and abound in the Pacific. British species are all small and less advanced than the tropical forms. They have adapted for a predatory life, and have an elongated and elaborate proboscis; the radula has been converted to a series of hollow teeth which can be charged with poison from a poison gland for killing the prey. None of these species occurs frequently and only one or two species of *Mangelia* can be collected alive between tide-marks.

Lora turricula (Montagu) (Fig. 99)

Shell turreted, 12–15 ribs, not prominent, on body whorl and those preceding it; ribs do not extend far below the periphery of the body whorl. All are crossed by numerous low spiral ridges which constitute the sole sculpture at the base of the body whorl. Outer lip angulated near top. Shell white. Dredged from sandy bottoms, commoner in north. 12 mm.

Philbertia teres (Forbes) (Fig. 108)

Shell a tapering cone with convex whorls marked by numerous spiral ridges, often alternately large and small; about 25 on body whorl, many fewer on the upper ones; mouth with large anal notch where it meets body whorl. Shell white with brown spots. Dredged from stony, shelly bottoms off northern coasts. Not common. 12 mm.

Philbertia gracilis (Montagu) (Fig. 106)

Shell recognizable by the long tapering cone with numerous ribs (about 14 on the body whorl) which do not quite reach to the suture at their apical end; mouth narrow, just less than half shell height; outer lip with well marked anal notch where it joins body whorl; siphonal canal long and opening out at its end. Shell fawn or brownish with light band at periphery. Dredged from muddy or sandy gravel off all coasts. Scarce. 30 mm.

Philbertia linearis (Montagu) (Fig. 107)

Shell with 12 ribs on body whorl crossed by as many spiral ridges; ribs extend to suture, which is rather deep as the whorls are distinctly convex; outer lip bulges outwards and has a slight anal notch; siphonal canal rather short. Shell cream, with brown lines. Dredged from coarse sand or sandy gravel off all coasts. 8 mm.

Mangelia nebula (Montagu) (Fig. 104)

Shell conical, with rather broad ribs, of which there are usually 11 on the body whorl; on each whorl they stop short of the suture, below which runs a ridge marked with small bead-like tubercles; whorls tumid; mouth rather narrow with distinct anal notch above; siphonal canal slopes to left. Shell dark brown, ribs paler. Can be found at LWST, but usually dredged on muddy gravel. 12 mm.

Mangelia powisiana (Dautzenberg) (Fig. 105)

Shell with a rather regularly conical spire, the whorls being flat-sided and not dipping to the sutures; below each suture is a slightly raised spiral ridge, largely plain, which prevents the broad ribs from extending to the sutures; 7–9 ribs on body whorl; many very fine striae visible; mouth narrow; siphonal canal short and wide, expanding a little at its end. Shell cream, yellow or brown. In sand at LW or dredged on muddy gravel on west coasts. 6 mm.

Mangelia attenuata (Montagu) (Fig. 100)

This species is usually recognizable by the rather sharp, narrow ribs, of which there are 9 on the body whorl; they extend to the suture, to which the whorls also dip; mouth long and narrow; siphonal canal straight. Shell brown, with 2–3 darker spiral bands, one below the periphery. Dredged off west coasts on muddy gravel; uncommon. 12 mm.

Mangelia costulata (Risso) (Fig. 103)

Shell marked by well developed ribs, 9 on body whorl, extending right up to sutures and almost to the base of the body whorl; these are covered with fine spiral striae; mouth narrow with marked anal notch on outer lip where it joins the body whorl; siphonal canal long. Shell pale brown. Dredged from gravelly bottoms in northern waters only; scarce. 12 mm.

Mangelia coarctata (Forbes) (Fig. 102)

Shell with 7 ribs on body whorl, which extend to the suture but not to the base, which is left very smooth; mouth narrow, with distinct anal notch where the slightly curved outer lip meets the body whorl; siphonal canal short and broad. Shell yellow-brown with some darker markings; mouth and the inside of the outer lip usually with stain of purple-brown. Dredged on sandy gravel off south and west coasts; locally not uncommon. 10 mm.

Mangelia brachystoma (Philippi) (Fig. 101)

Shell marked by broad ribs crossed by spiral ridges; 7–9 ribs on body whorl crossed by about 15 ridges; ribs flatten before reaching suture and disappear below periphery of body whorl, though spiral ridges persist; mouth narrow, a little expanded above, with small notch; siphonal canal short and wide. Shell yellow-brown or dark brown. Dredged from mud or muddy gravel off south and west coasts. 7 mm.

ACTEONIDAE

Shell spirally wound, with very large body whorl and small conical spire; columella with a fold which appears as a tooth on looking at the mouth; operculum horny. Mantle cavity opens anteriorly and to the right, on which side there is an exhalant siphon; the cavity also extends in the form of a caecum alongside the visceral mass to the apex of the shell. Ctenidium with triangular folded leaflets. Cephalic tentacles flat lamellae, their bases meeting in the mid-line. Visceral loop streptoneurous. Hermaphrodite. Larva a veliger.

This family is classified with the opisthobranchs, but is included here because with its thick, spirally wound shell, operculum and large uninvaginable penis it presents a great resemblance to the prosobranchs. There is only one British species, *Acteon tornatilis* (L.), sometimes called a barrel shell or little barrel, found on sandy beaches at LW or below. During the breeding season it extends much higher on the beach to lay its spawn masses. The animals burrow shallowly, are carnivores (probably eating small polychaete worms) and have a whitish shell with brown bands round it. It occurs on all coasts. 15 mm.

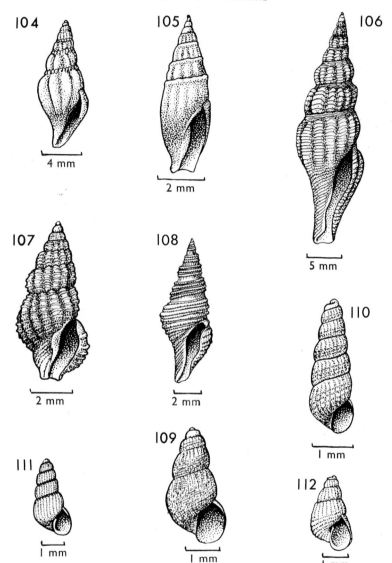

Fig. 104, *Mangelia nebula*; Fig. 105, *Mangelia powisiana*; Fig. 106, *Philbertia gracilis*; Fig. 107, *Philbertia linearis*; Fig. 108, *Philbertia teres*; Fig. 109, *Chrysallida decussata*; Fig. 110, *Chrysallida indistincta*; Fig. 111, *Chrysallida obtusa*; Fig. 112, *Chrysallida spiralis*.

PYRAMIDELLIDAE

Shell small, conical, with few turns, apex sinistral though rest of shell coils dextrally; columella with one or two teeth. A shelf-like projection (mentum) on underside of head, below mouth; eyes between tentacles which are grooved on outer side. No radula. Gut highly modified with stylet, long proboscis and buccal pump for parasitic life. Salivary glands tubular; no oesophageal gland. Ctenidium absent. Animals hermaphrodite. Larval state largely unknown. These animals are opisthobranchs.

1. Shell tall with more than 10 whorls .   . *Turbonilla elegantissima* (p. 106)

Shell short with less than 10 whorls **2**

2. Shell with ribs and spiral ridges, sometimes decussating; with or without tooth on columella . . . . *Chrysallida* (**a1**)

Shell smooth, sometimes with very slight keel on periphery of body whorl; tooth always present on columella . . . . . *Odostomia* (**b1**)

a1. Tooth on columella; ribs and spiral ridges do not interact to give reticulate appearance on penultimate whorl though some sign of this may be visible on body whorl **a2**

No tooth visible on columella; ribs and spiral ridges interact to give distinct reticulation on penultimate and body whorls **a3**

a2. Base of shell marked with spiral ridges; associated with *Sabellaria*
Chrysallida spiralis (p. 106)

Base of shell with longitudinal markings, or plain; associated with *Ostrea*
C. obtusa (p. 106)

a3. Shell tall, about 7 whorls; mouth ⋫ one-quarter shell height
C. indistincta (p. 106)

Shell squat about 4 whorls; mouth about two-fifths shell height
C. decussata (p. 106)

b1. Body whorl of shell slightly keeled (*this must be looked for carefully*); mouth squarish **b2**

Body whorl of shell not keeled mouth; round **b3**

b2. Spire comes to sharp point; tooth prominent; mouth angulated; colour of shell bluish-grey; colour of animal clear white; associated with *Pomatoceros*
Odostomia unidentata (p. 107)

Spire comes to blunt point; tooth rather weak; mouth round; colour of shell yellowish; colour of animal frosted white . . . *O. turrita* (p. 107)

b3. Tooth easily seen, but small; mouth round; associated with *Pomatoceros*
                                                                                    b4

   Tooth prominent; mouth elongated; associated with bivalves . . .   **b5**

b4. Outer lip curved outwards and joining body whorl almost at right angles; spire truncated; body whorl occupies rather more than half shell height; yellowish . . . . *O. lukisi* (p. 107)

   Outer lip not curved outwards, straight, and joining body whorl at acute angle; spire pointed; body whorl occupies about half shell height; brownish . . . . *O. plicata* (p. 107)

b5. Spire rather straight-sided, whorls not dipping to sutures; mouth not turning outwards at base; white with purplish tinge; on ears of *Pecten* or *Chlamys*
                                                                    O. eulimoides (p. 107)

   Spire shows whorls dipping distinctly to sutures; mouth turned outwards at base; yellowish; with *Mytilus* . . . . *O. scalaris* (p. 107)

The pyramidellids are a family highly modified for an ectoparasitic life on a variety of hosts (see Ankel, 1948; Fretter and Graham, 1949; Fretter, 1951*b*). They are opisthobranchs, but are included here because they are almost certain to be regarded as prosobranchs by collectors. They are extremely difficult to identify and attention must be paid to details of shell structure indicated above and shown in the figures. The source of the specimen is important and it should be noted what animal they are parasitizing. On occasion they may seek shelter away from their hosts when the tide is out: therefore search in their vicinity for a probable host.

Turbonilla elegantissima (Montagu) (Fig. 119)

Shell awl-shaped with 11–12 whorls, each marked with many low flexuous ribs, absent from the basal part of the body whorl; white and glossy. Animal white. Found at LWST and below in muddy situations where it parasitizes the polychaete worms *Audouinia* and *Amphitrite*. Locally common. 8 mm.

Chrysallida spiralis (Montagu) (Fig. 122)

Occurs at LWST and below on tubes of the polychaete *Sabellaria*. It is recognizable by the sharp separation on the body whorl of ribs at the apical end and spiral ridges basally. 3 mm.

Chrysallida obtusa (Brown) (Fig. 111)

Occurs predominantly with oysters, usually dredged. Similar to *C. spiralis* but lacks the basal spiral lines of that species. 3 mm.

Chrysallida indistincta (Montagu) (Fig. 110)

Recognizable by the tall shell and small mouth. Its relationship to other animals is not known. Commoner apparently in the north. 3–4 mm.

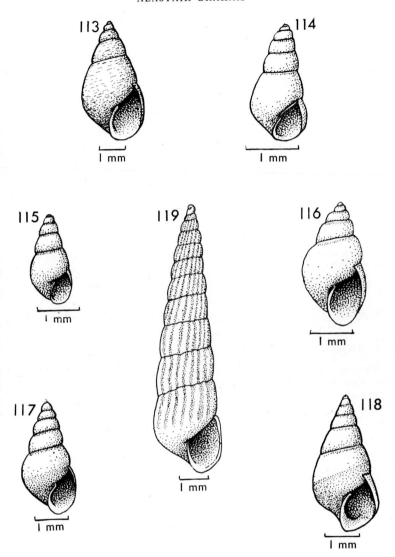

Fig. 113, *Odostomia eulimoides*; Fig. 114, *Odostomia lukisi*; Fig. 115, *Odostomia plicata*; Fig. 116, *Odostomia scalaris*; Fig. 117, *Odostomia turrita*; Fig. 118, *Odostomia unidentata*; Fig. 119, *Turbonilla elegantissima*.

Chrysallida decussata (Montagu) (Fig. 109)

The host is not known. Dredged. 3 mm.

Odostomia unidentata (Montagu) (Fig. 118)

Recognizable by rather straight-sided, bluish-grey, sharply pointed conica shell, with well marked columellar tooth. Common intertidally on stones mo⟩ or less completely covered with the polychaete *Pomatoceros*. 4 mm.

Odostomia lukisi Jeffreys (Fig. 114)

Found in similar habitats to *O. unidentata* in south, only dredged in nortl Distinguished from *unidentata* by blunt end to spire, convex whorls dipping t suture, small size of columellar tooth, yellowish colour and size. 2 mm.

Odostomia turrita Hanley (Fig. 117)

Host unknown, not found in south. Similar to *O. unidentata*, but has blun point to spire, weak tooth, narrow mouth and the body of the mollusc is speckle with white points. 3 mm.

Odostomia plicata (Montagu) (Fig. 115)

Occurs with *Pomatoceros*. Distinguished from *O. unidentata* by the small siz (2 mm) and lack of keel on body whorl; from *O. lukisi* by straightness of oute lip. At LWST and below in south, only sublittoral in north.

Odostomia eulimoides Hanley (Fig. 113)

Confined to the ears of scallop shells, on which it is not uncommon. 4 mm.

Odostomia scalaris Macgillivray (Fig. 116)

Common in mussel beds, more especially in north. 3 mm.

There are about two dozen further species in the family Pyramidellidae whic⟩ have been recorded in the British fauna. Their occurrence, however, is either s⟨ infrequent, or limited to dead shells, that it has not seemed right to include then here.

Literature

The previous pages contain references to about 140 species of gastropod: the British fauna (including animals found only in the Channel Islands) has over 230 species according to Winckworth (*Journal of Conchology*, 1932, vol. **19**, pp. 211–52). The difference between the two figures is represented by species which are so rarely found as not to deserve inclusion in a work of this type. If it is desired to make further investigations of the identity of an animal which cannot be run down from the keys given here, or to obtain more detail about one which has been identified or doubtfully identified, then recourse must be had to other volumes. Of these the two standard works are still (1) *A History of British Mollusca, and their Shells* by E. Forbes and S. Hanley, published 1849–53 in 4 volumes by Van Voorst, London. The text is difficult for modern readers in that the vocabulary is extremely technical, but the figures of shells and soft parts, published in volume 4, are useful. They should, however, be checked in certain critical cases against commentaries published by J. T. Marshall in the *Journal of Conchology*, Vols 7 (1893–4), **8** (1895), **9** (1898–9), **10** (1901–2), **13** (1910–12), **14** (1913–15), **15** (1916–18). (2) *British Conchology*, by J. G. Jeffreys, published 1862–9 in 5 volumes by Van Voorst, London. Jeffreys' book gives the most careful and critical verbal descriptions of the species though the illustrations are fewer and perhaps less clear than those in Forbes and Hanley.

After the publication of these works there is a long gap which is only now beginning to be filled. The freshwater and terrestrial operculates are dealt with in *British Snails* by A. E. Ellis (Oxford, Clarendon Press, 1926). The text gives adequate descriptions of the shell, often quoting the original account; the illustrations are photographs of the shells, which give good outlines but often fail to reveal sculpture satisfactorily. Rather better are the drawings in the Freshwater Biological Association's *Key to the British Fresh- and Brackish-Water Gastropods* (Scientific Publication No. 13, 3rd edition 1969) by T. T. Macan, which accompany a key annotated with anatomical and ecological notes. This is a very useful handbook, though it deals with only 14 operculate species. A recent volume (*British Shells*, by Nora F. McMillan; London, F. Warne and Co., 1968) deals with most of the species of gastropod, operculate and otherwise. Figures of 200 species are given and these are good.

The above books deal primarily with identification. A few others may be mentioned as dealing with some aspects of the life of the animals:

(1) an excellent introductory volume: *Molluscs*, by J. E. Morton; Hutchinson University Library, London, 1967;
(2) a general account of prosobranchs: *British Prosobranch Molluscs*, by V. Fretter and A. Graham; London, The Ray Society, 1962;
(3) an account of some special aspects such as feeding, digestion and distribution: *The Biology of the Mollusca*, by R. D. Purchon; Oxford, Pergamon Press, 1968.

References

ANKEL, W. E. 1948. Die Nahrungsaufnahme der Pyramidelliden. *Verh. dt. zool. Ges.*, *Zool. Anz.* Suppl. **13**, 478–84.

BARKMAN, J. J. 1955. On the distribution and ecology of *Littorina obtusata* (L.) and its subspecific units. *Arch. néerl. Zool.* **11**, 22–86.

DAUTZENBERG, P. and FISCHER, H. 1914. Etude sur le *Littorina obtusata* et ses variations. *J. Conch., Paris*, **62**, 87–128.

FORBES, E. and HANLEY, S. 1849–53. *A History of British Mollusca, and their Shells.* 4 Vols. **2** (1849), **3** (1850), **4** (1852, 1853). London, van Voorst.

FRETTER, V. 1948. The structure and life history of some minute prosobranchs of rock pools: *Skeneopsis planorbis* (Fabricius), *Omalogyra atomus* (Philippi), *Rissoella diaphana* (Alder) and *Rissoella opalina* (Jeffreys). *J. mar. biol. Ass. U.K.* **27**, 597–632.

FRETTER, V. 1951a. Observations on the life history and functional morphology of *Cerithiopsis tubercularis* (Montagu) and *Triphora perversa* (L.). *J. mar. biol. Ass. U.K.* **29**, 567–86.

FRETTER, V. 1951b. *Turbonilla elegantissima* (Montagu), a parasitic opisthobranch. *J. mar. biol. Ass. U.K.* **30**, 37–47.

FRETTER, V. 1955. Observations on *Balcis devians* (Monterosato) and *Balcis alba* (da Costa). *Proc. malac. Soc. Lond.* **31**, 137–44.

FRETTER, V. and GRAHAM, A. 1949. The structure and mode of life of the Pyramidellidae, parasitic opisthobranchs. *J. mar. biol. Ass. U.K.* **28**, 493–532.

FRETTER, V. and GRAHAM, A. 1962. *British Prosobranch Molluscs.* London, Ray Society.

FRETTER, V. and PATIL, A. M. 1958. A revision of the systematic position of the prosobranch gastropod *Cingulopsis* (= *Cingula*) *fulgida* (J. Adams). *Proc. malac. Soc. Lond.* **33**, 114–26.

GRAHAM, A. 1938. On a ciliary process of food-collecting in the gastropod *Turritella communis* Risso. *Proc. zool. Soc. Lond., A*, **108**, 453–63.

JAMES, B. L. 1968a. The characters and distribution of the subspecies and varieties of *Littorina saxatilis* (Olivi, 1792) in Britain. *Cah. Biol. mar.* **9**, 143–65.

JAMES, B. L. 1968b. The distribution and keys of species in the family Littorinidae and of their digenean parasites, in the region of Dale, Pembrokeshire. *Field Studies*, **2**, 615–50.

JEFFREYS, J. G. 1862–9. *British Conchology.* Vols. **1–5.** **1** (1862), **2** (1863), **3** (1865), **4** (1867), **5** (1869). London, van Voorst.

LAURSEN, D. 1953. The genus *Ianthina. Dana-Report* no. **38**, 1–40.

LILLY, M. M. 1953. The mode of life and the structure and functioning of the reproductive ducts of *Bithynia tentaculata* (L.) *Proc. malac. Soc. Lond.* **30**, 87–110.

McMILLAN, N. F. 1939. The British species of *Lamellaria. J. Conch., Lond.* **21**, 170–3.

McMILLAN, N. F. 1968. *British Shells.* London, F. Warne.

WILSON, D. P. and WILSON, M. A. 1956. A contribution to the biology of *Ianthina janthina* (L.). *J. mar. biol. Ass. U.K.* **35**, 291–305.

YONGE, C. M. 1937. The biology of *Aporrhais pes-pelicani* (L.) and *A. serresiana* (Mich.). *J. mar. biol. Ass. U.K.* **21**, 687–704.

YONGE, C. M. 1938. Evolution of ciliary feeding in the Prosobranchia, with an account of feeding in *Capulus ungaricus. J. mar. biol. Ass. U.K.* **22**, 453–68.

YONGE, C. M. 1946. On the habits of *Turritella communis* Risso. *J. mar. biol. Ass. U.K.* **26**, 377–80.

Index to Species

Figures in *italics* refer to illustrations.